A Guide to ~~Oral and Maxeo...~~ ~~Maller...~~The Guy or Gal Who Takes Your Teeth Out

A Guide to ~~Oral and Maxeo...~~ ~~Maller...~~The Guy or Gal Who Takes Your Teeth Out

A Young Person's Guide: From Extractions to Implants

Donald R. Grippo, D.D.S.

2007

A Guide to ~~Oral and Maxeo...~~ ~~Maller...~~The Guy or Gal Who Takes Your Teeth Out

Pauline- you are my best friend, my advisor, my strength, and my source of inspiration.

Rom, Jess, and Kris-as E.E. Cummings wrote, "I carry your heart with me (I carry it in my heart).

Table of Contents

Knowledge is power
…..Sir Francis Bacon

Forewarned, forearmed; to be prepared is half the victory
…..Miguel de Cervantes

Introduction

One late winter afternoon, a precocious eight-year-old boy came to my office to have a tooth removed. It was a baby tooth and, from my point of view, a relatively easy tooth extraction. But the young man with spiky blond hair sitting in my dental chair produced from his pocket a wrinkled piece of paper. On it, he had written several questions.

With the poise of one much older, he proceeded to ask me those questions. They were clear and concise. He wanted to know what was wrong with his tooth and how I was going to fix it. Of course, I answered his questions truthfully and in a straightforward manner. He seemed to demand such honesty. I eventually extracted his tooth, and he went home a happy camper.

At the end of that day, when my helpers had gone and the office was finally quiet, I sat reflecting in my private office. How wonderful it would have been if, prior to coming to me, the child could have read about the experience he might expect to have. Not in the First Visit to the Dentist type of book, but in a book that realistically dealt with oral and maxillofacial subjects and provided real-world information.

Why hadn't I thought of it sooner? Hadn't I been practicing oral and maxillofacial surgery for more than thirty years? I gave myself a mental kick in the butt, and decided to write informative short stories for the Randys and Karens who need the services of people of my profession. These stories would let the reader know what to expect when visiting the oral and maxillofacial surgeon, so that they would not be frightened.

I feel lucky to have picked oral and maxillofacial surgery as my career and to have practiced it all these years. What a great field of endeavor. Oral and maxillofacial surgeons bridge the gap between dentistry and medicine. We get to do all sorts of interesting things. For example, we treat facial injuries from accidents of all types. The correction of cosmetic and functional facial deformities is also an important part of my profession. Taking out wisdom teeth, removing tumors, and cysts, giving intravenous anesthesia, and placing dental implants are all part of the challenging daily routine of our practices. The profession is continually changing, as new techniques, materials, and training dictate.

Yet the average person thinks of us as tooth pullers. It seems that most of my patients can't pronounce maxillofacial, much less know what it means (i.e., relating to the facial bones and associated structures). They would be surprised to learn that many of my colleagues perform hair transplantation, laser skin resurfacing, and complicated neck surgery. Our professional name says a great deal of what we do, but doesn't reveal everything we do. Perhaps in the future we'll change our professional name, as we've done in the past, to better reflect the scope of our surgery.

We began with Hippocrates, the father of medicine who was, also, the first oral and maxillofacial surgeon. In ancient Greece, he performed procedures such as extracting teeth and treating facial injuries. In the more recent past, we were called exodontists, which referred to the fact that pulling teeth was our major job. Next, we became oral surgeons, indicating that we did much more than remove teeth. The problem with this name was that it seemed to limit us to the inside of the mouth. The mouth is a nice place, but we were working in other areas as well. We needed a name that said our field included all the facial bones; hence, the name oral and maxillofacial surgery.

Even oral and maxillofacial surgery doesn't fully define the scope of our profession. We are expanding into more areas of cosmetic and reconstructive surgery. New tools and techniques are allowing us to add to our surgical repertoire. Should we have a new name? Perhaps, but a new name is hard to find. It might have to include the words cosmetic, plastic, facial, reconstructive, head and neck, and implantology. Are we oral, maxillofacial, cosmetic, plastic, facial, reconstructive, head and neck, implant surgeons? We truly are, but that name is much too long. Our present name is long and misunderstood enough as it is. Also, other professions, such as the facial plastic surgeons, use these names, which add to the confusion. So it seems that for the time being, we are oral and maxillofacial surgeons.

My stories illustrate oral and maxillofacial problems faced by children and young adults. It is my hope that reading them will not only supply the reader with detailed and accurate information regarding their treatment, but will also help allay some of the fears associated with the dental office. As a result, I won't hear what I hear much too often from patients: "I can't tell you how long I've dreaded seeing you."

Molly

I

When Molly Diaz woke up this morning, her first thoughts were: Tomorrow is July 17th, my birthday. I'll be twelve years old and only one more year until I'm a teenager. I can hardly wait until then. I also can't wait for the birthday party Mom has planned for me. Molly's room was bright with the morning sun shining through her curtained window. She had on her lucky pajamas, which meant that today was going to be a great day.

She loved the birthday parties that Mom gave. They made her feel very special. Molly was having a wonderful summer. She had spent two weeks at swim camp, where she finally mastered several synchronized swimming techniques. And she was looking forward to the family trip to Cape Cod the first week of August. Molly loved school, though, and wouldn't mind going back in September. But for now, she was having fun.

As she hopped out of bed, something was bothering her. Molly felt pain in a tooth that had recently developed a hole in it. It wasn't a great deal of pain, and she was hoping it would go away. She didn't want a toothache spoiling her birthday party. Molly had noticed the hole in one of her lower left teeth a few days ago, but it hadn't hurt this much before.

Mrs. Diaz was downstairs making breakfast, and the smell of bacon that wafted into the room temporarily took her mind off her tooth. Molly had the habit of putting the clothes she intended to wear next to her bed the night before. It made dressing easy and fast. Today, it was cutoff jeans and a yellow T-shirt. She brushed her teeth, and the tooth seemed okay. She loved the smell and taste of bacon, and quickly went downstairs to eat.

When Molly came to the kitchen table, Mrs. Diaz turned and looked closely at her. "Molly," she said, "your face looks a bit swollen on your left side. How do you feel, dear?"

"Fine, Mom." Molly didn't want Mom to know that her mouth hurt, and she was still hoping the pain would go away. Her dad, a stockbroker for the firm of Paul and Stanley, had already eaten and gone to work. She sat down to breakfast. The big round table was set perfectly, as always. Along with the bacon were eggs, toast, milk and orange juice. A brightly colored cloth covered the table. She picked up her glass of juice and took a sip.

"OWWW!"

"Molly, what's wrong?"

"My tooth! It hurts! The pain goes all the way to my ear!" Molly was close to tears.

"How long has it been hurting?"

"For a few days."

In a worried voice, "Why didn't you tell me about it?"

"It didn't hurt a lot, and I was hoping it would go away."

"I'm calling Dr. Chen now!" Mrs. Diaz said with a look of concern. She picked up her cell phone and dialed Dr. Chen's number that she had gotten from a card on the refrigerator. She went to her study where she had a writing desk.

Dr. Cynthia Chen treated only children, and Molly liked her very much. Visits to her office were fun, as the reception room was wildly decorated and had video games for children to amuse themselves. And there were always children of Molly's age to keep her company. Also, Dr. Chen had never hurt her.

Mrs. Diaz came back to the kitchen. "Dr. Chen checked your chart and feels that if you have a cavity that's giving you pain, your tooth may need to be taken out. She recommended that I take you to Dr. Giles, a specialist in taking out teeth. I made an appointment for him to see you this morning. We were lucky, there was a cancellation and they will see you as soon as we can get there."

Molly was nervous. She knew and liked Dr. Chen, but she didn't know Dr. Giles. He might hurt her. "Mom, my tooth feels much better. I don't think I need it pulled. I don't want to go to Dr. Giles's office today. Besides, Dad would want to come with us. We should wait for him to come home from work." Molly was using all the excuses she could think of to postpone going to this unknown dentist.

Mrs. Diaz looked at Molly. "I know how you feel, but Dad would agree it's important that we take care of your tooth before it becomes a big problem. I'm sure you'll like Dr. Giles." Mrs. Diaz told Molly to get dressed in something more appropriate than what she had on.

The pain in the tooth didn't stop Molly from carefully brushing her hair and picking out clothes that she thought made her look good. After all, boys—in whom Molly had only recently begun to take an interest—might be in the waiting room. When she was ready, she and her mom drove to the dentist's office. At the moment, she didn't think her lucky pajamas were so lucky.

II

Dragging slowly behind her mother, Molly entered Dr. Giles's office. It was a gorgeous day. It was mostly sunny with a few clouds scattered about, and she had played "find the faces in the clouds" on the way to the office. She had only found three faces by the time they had arrived, but the game kept her from thinking about the fact that she was going to the dentist. The sign in front read Gordon Giles, D.D.S., Oral and Maxillofacial Surgery, which Mom had read out loud for her. The title, Oral and Maxillofacial Surgery, made Molly even more nervous. Why did a dentist need such a long name just to take a tooth out? In the reception room, Molly noticed that there were people of all ages waiting to see the doctor. She would have preferred kids only.

Her mom went to the receptionist's window. Molly noted that her mom looked good. She always did when she went out of the house. Molly probably got her own habit of wanting to look her best at all times from her mother. "Mrs. Diaz with Molly to see Dr. Giles," Mrs. Diaz said, smiling warmly.

The receptionist smiled back. "Take a seat, and please fill out these forms. The doctor will see Molly shortly."

They took seats. Molly was quiet while Mrs. Diaz was filling out the forms. The room seemed awfully warm, and Molly was sweating a bit. In the chair to Molly's right sat the only other child in the room. She thought that the man sitting on the other side of the boy was probably his father. The boy looked a little older than Molly, but not much. He was also very good-looking. Molly was glad she had taken the time to look her best before leaving home. She wanted to ask him why he was there, but she was naturally shy and didn't want to speak first. (This was not the case when she was with a bunch of girls in her bedroom on a Friday night pajama party.) Molly casually turned her head toward the boy, as if she were just looking around the room. At the same time, the boy casually turned his head toward Molly, as if he, too, were just looking around the room. Their eyes met.

"Hi," said the boy.

"Hi," said Molly.

"What's your name?"

"Molly. What's yours?"

"Ben. Why are you here, Molly?"

"I *may* need to have a tooth taken out." Molly, of course, was secretly hoping that she might not need the tooth out.

"Me, too. Only I need two out."

"Two! Wow!"

"No big deal. I had two other teeth out two weeks ago, and it was easy."

"Four teeth! How come?"

"My orthodontist says I have too many teeth in my mouth, and I need some taken out so he can straighten them," said Ben. When he spoke, he was quite animated and gestured with his hands. "See where I had the two teeth taken out?" Ben opened his mouth and pulled his cheek back with his finger, showing the spaces where the teeth had been recently removed.

"Won't the spaces show?"

"No, my orthodontist says that after he is finished with me, no one will know I had any teeth taken out."

"Wow," said Molly again. "Did it hurt when they were pulled?" Molly very much wanted to know the answer to this question, although she tried not to show it.

"Not a bit. The doctor put my teeth to sleep, and I only felt a little pressure when they came out."

Molly was relieved by Ben's answer. She liked him, and the room seemed not as warm as before.

III

A nurse in a pretty blue shirt and pants opened a door to the reception room. "Molly," she said, looking at Molly and Mrs. Diaz, "I'm Doris. I'll be helping Dr. Giles today. Please follow me." Doris waited while Mrs. Diaz and Molly approached. She then placed an arm around Molly and with a smile that went from ear to ear, she said, "Welcome to our office, Molly." Molly had a good feeling about Doris.

Molly and Mrs. Diaz walked with Doris down a carpeted hall. Paintings of sailboats were on the walls. Molly slowed a bit to look at them. One painting showed a sailboat in a dark green sea, tipping almost on its side because of the wind. In another, a sailboat was in a storm. In this painting the sea was almost black, except for the white spray on top of the waves. In the largest painting, several sailboats were sailing in a blue horizon on a beautiful day. It was obvious to Molly that at least one person in the office loved sailboats.

They entered a room that had a chair similar to the chair in Dr. Chen's office. But it had other equipment in it that she had not seen before. A strange-looking gray machine in the corner had hoses and a balloon coming out of the bottom. Also, there was no light hanging from the ceiling as in Dr. Chen's office.

Doris seemed awfully friendly and had a nice smile. As a result of Ben's reassurance and Doris' friendliness, Molly was much more relaxed than when she first entered the office.

"Molly, please sit in this chair so we can take an x-ray." Doris indicated the chair Molly had previously noticed. It was light blue in color, and looked like a great chair to sit in while watching TV.

"Mrs. Diaz, I'll have you stand outside the room while I take an x-ray, and then you can sit in that chair over there." Doris pointed to a chair near the entrance of the room.

Molly sat in the assigned chair, and Doris placed a thick drape over Molly's chest and neck. Doris explained the purpose of the drape, but as a result of her experience in Dr. Chen's office, Molly already knew the drape protected her while the x-ray was taken. Next, Molly opened her mouth as Doris directed. She also knew what x-rays were. Dr. Chen's assistant had explained that they were pictures that allowed the dentist to see inside the jawbone and teeth in order to find any problems with them.

Doris bent over her and put a little square of plastic in Molly's mouth, and Molly bit down on it. Molly noticed that

Doris was wearing a delightful perfume. She thought she might later ask her its name. What looked like a robot's arm, which was attached to the wall, was placed next to her face. It made a chirp like a cricket when the x-ray was taken. It didn't hurt.

Molly's mom came back into the room and sat in the previously assigned chair.

"I'll be right back after I develop this x-ray." Doris closed the door, leaving Molly and Mrs. Diaz alone. Except for soft background music, the room was rather quiet with the door closed. Molly momentarily had the feeling that she had when she entered a library, that is, she should speak softly, or not at all.

Molly continued her inventory of the room. She noticed that there was a stand with a tray on it. The tray was covered with a green cloth, but by the lumps and bumps in the cloth she knew that the doctor's tools were under it. Molly recognized a tube coming out of a jar that was attached to the wall on her left. The tube had a silver tip that was in a holder, which was also attached to the wall. Dr. Chen had the same thing in her office. It was called a spit straw and was used to keep her mouth dry by taking extra saliva away. There was a tall pole in one corner that had a bag of what looked like water hanging from it. The bag had a long skinny tube sticking out of the bottom. There were other pieces of equipment in the room, but she wasn't sure what they were. She continued to sit quietly, and Mrs. Diaz did the same.

In a few minutes, Doris came back into the room and hung the little x-ray on a glass box that had a light in it. Doris looked at Molly and Mrs. Diaz and said, "There's a big hole in your baby molar tooth, with an abscess at the bottom of the roots. That tooth will have to be taken out so that the permanent tooth below it isn't damaged. I'm sure Dr. Giles will say the same thing." Seeing the concern on Molly's face, Doris continued, "Don't worry, Molly, this isn't going to hurt you." Molly let out a deep breath of relief.

Doris turned toward Mrs. Diaz and said, "Dr. Giles will be here in a moment, to review Molly's health history and to talk about the treatment that she needs."

IV

A short while later the doctor entered the room. He had on a shirt and pants similar to those Doris wore, but his were green. He also had a green hat that doctors wear, tied in a bow in the back. He looked a little like Molly's uncle Carlos.

"Hi, I'm Dr. Giles," he said to Molly. He walked over to Mrs. Diaz, who stood up, and shook her hand. Dr. Giles then sat on a stool with wheels and rolled over to Molly, while Mrs. Diaz sat down again.

He was a tall man with gray hair sticking out along the edges of his cap. His face was long like his body, and he looked very friendly. Molly instinctively trusted him. Dr. Giles took the papers that Doris handed him and began to look through them. "Let's see," he said, as he looked toward Mrs. Diaz. "It says here that Molly is in good health."

Mrs. Diaz nodded her head in agreement.

Dr. Giles had dark brown eyes, and they looked directly at Mrs. Diaz as he spoke. "There are no allergies to medicines, and she doesn't take any on a daily basis. It looks like she has no medical problems. Taking out a tooth should be a piece of cake."

Molly didn't like the idea of having her tooth out, but she liked the "piece of cake" part.

Dr. Giles rolled his chair to the lighted box that the x-ray was on. Doris gestured to Mrs. Diaz to stand next to Dr. Giles, so she would have a good view of the x-ray. He confirmed what Doris had said. "The x-ray shows a big cavity in your baby tooth, Molly, and an abscess at its roots. This baby tooth is called the second primary molar tooth. By Molly's age, many children have already lost this tooth." Dr. Giles pointed to the cavity with his pen. "It's important to extract the baby tooth so that there's no damage to the permanent tooth underneath it.

You can see the permanent tooth here on the x-ray, just below the baby tooth." He pointed to the permanent tooth. "It will soon take the place of the baby tooth. We call it the second bicuspid tooth. Infection from the baby tooth can seep down and damage it." He looked at Mrs. Diaz and then to Molly.

Dr. Giles rolled back to Molly. He put a light on his head that looked like a coal miner's light. He took his small mirror and proceeded to look in Molly's mouth. To Molly, it felt as if he had a lollypop in her mouth. "Mrs. Diaz, I'd like you to have a peek at Molly." Mrs. Diaz walked over to Dr. Giles and looked over his shoulder. "Molly, would you like a hand mirror, so you can see too?"

"No, Dr. Giles," Molly answered softly. She preferred not to see what was going on in her mouth.

Mrs. Diaz looked into Molly's mouth while Dr. Giles shined his light on the offending tooth. He gently pulled out Molly's cheek with his little mirror so Mrs. Diaz could get a better view of the tooth. "The decayed tooth is the last baby tooth remaining in Molly's mouth." Dr. Giles pointed to the tooth with a small instrument, so Mrs. Diaz knew to which tooth he was referring. "You can see a permanent tooth right behind it. That's called the first molar tooth." He pointed to the first molar tooth. "Often, when this baby tooth is taken out, we have to keep the space open until the permanent tooth underneath it comes in. This is done so the first molar tooth doesn't move forward and close the space. If that happens, the tooth underneath the baby tooth may be blocked from coming in. In Molly's case, I expect it will quickly erupt, avoiding a space problem."

That's good news, thought Molly.

Dr. Giles paused a moment and then continued. "Notice the swelling of the gum tissue around the decayed tooth. It's responsible for the minor facial swelling that Molly has."

I hope that's not bad news, thought Molly.

"Such swelling is common," said Dr. Giles.

It's not bad news, thought Molly.

Looking at Molly, Dr. Giles asked, "Do you have any pain in the tooth?"

Molly was tempted to say no, but knew that she should be truthful. "Yes, and it also goes to my ear."

Dr. Giles rolled his stool back a bit. It made a squeaking noise. "That's because of the way the nerve to the teeth runs. It runs next to the ear, causing the ear to hurt. Many people think they have an ear infection and go to their ear doctor first. When no ear infection is found, they're sent to the dentist. Usually, there is a bad tooth causing the pain." Do either of you have any questions?"

"No," said Mrs. Diaz.

"Yes," said Molly.

"What's your question, Molly?"

"What does oro-maxeo-face...I can't even say it, but what does it mean?"

Dr. Giles smiled. "Molly, many people can't say oral and maxillofacial surgery. And people are always asking me what it means. It means that I do much more than take out teeth. For instance, I treat people who injure their face in auto accidents. I correct problems, or what we call deformities, of the face. I also take care of people who have much more serious problems in their mouth than a simple infected tooth. The bottom line is that I treat people from the neck up. Good question, Molly. Do you have any others?"

"No, Doctor."

Molly was impressed with what Dr. Giles could do.

V

"Very well," said Dr. Giles. "The next thing we need to talk about is how we're going to take the tooth out, that is, what type of anesthesia we're going to use."

Molly gave Dr. Giles her undivided attention.

"What are our options, Doctor?" asked Mrs. Diaz.

"Well, Mrs. Diaz, we have four choices. The first choice is to take the tooth out with local anesthesia, or what is commonly called Novocaine®. With this technique, we first place a numbing cream on the gum. Then, an injection that puts the tooth to sleep is given. The injection feels like a little pinch. Once the tooth is asleep, it is removed painlessly. This is the simplest and quickest way to get the tooth out."

Dr. Giles looked at Molly. "The only thing that you would feel is pressure, like this." Dr. Giles placed his hand on Molly's shoulder and gave a gentle squeeze. "And you might hear a little noise."

Dr. Giles rolled back a bit to better speak to both Mrs. Diaz and Molly. "An interesting thing about local anesthesia is that it makes you think that your face is swollen out to here." At this point, he placed his hand a few inches from his lower lip to give the impression that it was swollen. "It isn't. Your face is really normal. It's just playing tricks on you."

"What are the other choices?" asked Molly.

"Choice number two is a combination of laughing gas and local anesthesia." Dr. Giles was now looking directly at Molly.

"Laughing gas," Molly repeated. That sounded like it might be fun. Anything that could make her laugh was fine with her.

"Tell us about laughing gas, Doctor," said Mrs. Diaz.

Dr. Giles continued. "Laughing gas, or what we call nitrous oxide, is given to you by a small breathing mask." Indicating with his finger, "That machine in the corner is used to give the laughing gas."

Molly looked there and saw that it was the machine with the hoses and balloon.

"We put the tiny mask over your nose and ask you to breathe through it. You breathe in the odorless gas, and soon you feel a slight tingling sensation in your body. That means the gas is working. Some people laugh when we give them nitrous oxide, that's why we call it laughing gas. It works by making you feel less pain. Local anesthesia is still used, but you don't feel it when I give it to you."

Molly looked at her mom with an expression that said laughing gas was her choice.

But Mrs. Diaz wanted to be sure she was aware of all the options. "What are the other two options, Doctor?"

"The third option is intravenous sedative anesthesia. To give this type of anesthesia, we start an intravenous line—that is, a small needle called a catheter is placed in a vein. It allows us to give fluids and medicines. This is the most effective way to give medicines, and the patient—Molly in this case—becomes very relaxed, or sedated. There's just a little pinch when starting the intravenous line. The bag of fluid you see hanging on that pole is the intravenous fluid, which is basically sugar water with a little salt and other minerals in it."

Dr. Giles paused to take a break and to allow Mrs. Diaz and Molly absorb what he was saying. After a few moments, he continued. "The fourth option is general anesthesia, in which the patient is completely asleep. This anesthesia is given by increasing the intravenous medications, by giving an anesthetic gas to breathe, or, sometimes, by an injection directly into the muscle of the leg, hip, or arm. These last two forms of anesthesia require that the patient go without eating or drinking for several hours before the appointment."

Dr. Giles paused again and turned toward Mrs. Diaz. His arms were folded across his chest. "Those are your choices, Mrs. Diaz. Molly and you should pick the one you feel is best."

"I'd prefer that we use local anesthesia or the laughing gas with local anesthesia, if you think they would work," stated Mrs. Diaz. "I feel that both Molly and I would rather not use the intravenous sedation or general anesthesia."

Molly nodded her head in agreement.

Dr. Giles also nodded and said, "I agree. In Molly's situation, they would work just fine. There are times when more profound anesthesia is necessary, but not in this case. I suggest that we ask Molly what she prefers." Dr. Giles looked at Molly. "What would you like?"

Molly thought for a moment and said, "I would like the local anesthesia with the laughing gas." To her, picking the anesthesia was as easy as choosing dessert. It was almost like saying, "I would like the ice cream with chocolate sauce instead of the cookies."

"Very well. If it's okay with you, Mrs. Diaz, I'll give Molly laughing gas and local anesthesia after the laughing gas starts to work."

"Are there any possible side effects I should know about?" asked Mrs. Diaz.

Dr. Giles smiled. "I'm glad you asked that question, Mrs. Diaz. I was just coming to that. You picked laughing gas and local anesthesia as Molly's anesthesia. Fortunately, while all anesthesia is potentially risky, your anesthesia choice is usually without complications. In rare instances, a patient may experience a bit of an upset tummy, but that doesn't last long. Laughing gas is a mild form of anesthesia, and problems, such as overdosing, are rare." He turned toward Molly and smiled again. "Safety is my most important concern, and I'm going to take all precautions to ensure a safe and pleasant experience for Molly." Turning to Mrs. Diaz, "Have I answered your question?"

"Yes, thank you, Doctor." Mrs. Diaz was visibly relieved.

VI

Doris spoke to Mrs. Diaz. "Now that we've reviewed the options available to take the tooth out and you've chosen one, I'm going to need your signature on our permission for extraction form. This form states that the treatment plan has been explained to you, and you understand and agree with it. Please read it over before you sign it." She handed the form to Mrs. Diaz on a clipboard with a pen. Mrs. Diaz read the permission form.

"It seems in order." She signed it and handed it back to Doris.

Doris continued, "Mrs. Diaz, would you like to be in the room while we treat Molly? The decision is up to you and Molly."

Mrs. Diaz looked at Molly. "What would you prefer?"

"Would you stay, Mom?" Molly looked at her mom with the best "pity me" expression she could muster. She knew her mother would do what she asked.

"Of course I will." Mrs. Diaz went back to her chair.

"Molly, Doris and I are going to put on our work clothes, and then we'll proceed with taking out your tooth," said Dr. Giles.

Doris put on a blue gown that wrapped around her. It appeared to be made of paper. She then helped the doctor put his on.

Molly asked, "They look like they're made of paper. Are they?"

"Yes, Molly, you're quite right. They're made of a special type of paper that we can throw away after we take out your tooth. It protects you and us from infection," said Doris.

Dr. Giles said, "I'm going to adjust the chair so you'll be lying back a bit." The chair automatically adjusted when Dr. Giles pressed a button on the back of the chair.

Dr. Giles stood on Molly's right, and Doris positioned herself on Molly's left. They were both standing, which was different from Dr. Chen's office. In that office, the doctor and her helper both sat, and Molly was placed in more of a lying down position.

Doris said, "I'm going to put a clip on your finger that helps us to see how well you are doing by measuring the oxygen inside of you. It doesn't hurt." After the clip was placed on her finger, Dr. Giles and Doris looked at a machine with red numbers to confirm that Molly was doing well.

Doris continued, "I'm now going to place this little mask over your nose, and have you breathe in and out slowly through your nose. You may smell the mint that I placed on it."

The mint smelled pleasant, and Molly began to breathe through her nose as directed. Soon she noticed a tingling in her hands and feet. She was feeling very good *and* very happy.

Doris told her that a numbing cream was being placed on the gum and local anesthesia was being given.

Molly could hear Doris, but her voice was far away and fuzzy. She also had a feeling of floating. She felt only a slight pressure in her mouth. After a while, she stopped floating and voices became normal.

Doris was telling her to close her mouth and bite on cotton. She was also telling her to continue to breathe in and out so that Doris could remove the nose mask.

Dr. Giles saw that Molly was doing well. He said to Mrs. Diaz that Doris would see that Molly left the office in good condition. Mrs. Diaz was to call him if there were any problems at home. Waving goodbye, he left the room.

VII

Within a few minutes Molly felt normal all over, except for the left side of her face. It felt swollen. She put her hand to her face, and she couldn't feel her lip.

Doris noticed Molly's actions and spoke to both Molly and Mrs. Diaz. "Molly's face is numb. As Dr. Giles said, it feels swollen, but isn't. The numbness will wear off in an hour or so. In the meantime, it's important that Molly not bite her lip to see how numb it is. It won't hurt now, but it will hurt when the feeling returns. And she could get an infection in it."

Doris continued "Molly is biting on cotton, and I want her to do so for another forty-five minutes. I'll give you more cotton to take home, so you can change it as necessary until all bleeding has stopped. Dr. Giles put a dissolvable medicine in the tooth socket to ensure that Molly will have no bleeding problems. You don't have to be concerned about that medicine, as it will dissolve within a few days." Doris paused to give Mrs. Diaz a chance to ask questions. There were none.

She handed Mrs. Diaz a pamphlet, "Here are written postoperative instructions, that are mostly common sense," Doris explained. "For example, Molly is to have a soft diet for forty-eight hours, with no hot foods. We don't care what she eats as long as her diet is nutritious. You can give her acetaminophen that is appropriate for her age if she has mild pain. If she has severe pain, which is unlikely, call Dr. Giles. Don't give her aspirin for pain, as it can result in prolonged bleeding and in rare cases, serious problems in children of her age. Molly needs to take it easy for a day or so, but in a few days she should be feeling fine. Molly, can you stand up?"

Molly felt fine, and was able to get up from the dental chair and stand without a problem.

"Here's your tooth, in case you want to put it under your pillow, Molly." Doris handed her a small plastic bag that had the extracted tooth in it.

Addressing Mrs. Diaz, she said, "Put the tooth in bleach for twenty minutes and then wash it with water. That will kill any germs on the tooth."

Doris then accompanied Molly and her mom back to the reception room. Molly noticed that Ben wasn't there, and hoped that he had as good an experience as she did. Doris said goodbye, and Molly and Mrs. Diaz left the office.

The sun was heating up the day, but a gentle breeze made it quite pleasant to be outside. "Thanks for bringing me to Dr. Giles, Mom. It was a lot easier than I thought it would be. Dr. Giles and Doris were very nice, and they didn't hurt me a bit." Molly also thought that it would be nice to meet Ben again. He would be a good friend to have.

"I'm glad you had a good experience." Mrs. Diaz replied. "If you ever need another tooth taken out, you won't be afraid to have it done."

Molly nodded and thought: I guess my pajamas were lucky after all.

She looked up at her mom. "And Mom, don't forget the two best things."

Mrs. Diaz looked puzzled. "What two things, Molly?"

"Don't forget my milk shake on the way home and my birthday party tomorrow."

Both Mrs. Diaz and Molly laughed as they walked to their car.

Definitions: Molly

Novocaine® (Local Anesthesia): Novocaine® is no longer used in dentistry. It was the first medicine that we call a local anesthesia, which was made to put teeth to sleep by an injection. Today, newer and better medicines such as lidocaine or mepivicaine are used. Dr. Giles and other dentists continue to speak of Novocaine® because most people are familiar with the word and understand what it means. In reality, to speak of Novocaine® when talking about local anesthesia is the same as speaking of George Washington when talking about presidents.

Nitrous Oxide (Laughing Gas): When we breathe in nitrous oxide, we feel less pain. Many people laugh when this gas is given to them, which is the reason it is also called laughing gas. Molly thought getting laughing gas would be fun. In 1844 Dr. Horace Wells, a dentist in Hartford, Connecticut, was the first dentist to use nitrous oxide in his office. He had the gas given to him while another dentist took out one of his teeth. He felt no pain. Since that time, dentists and physicians have been using nitrous oxide when performing surgery.

Pulse Oximeter: The clip that Doris placed on Molly's finger was part of a machine called a pulse oximeter. This machine continuously measured the amount of oxygen in Molly's body. By watching the numbers that appeared on the machine, Dr. Giles could tell if Molly was breathing properly.

X-rays: X-rays are invisible rays made by an x-ray machine. These rays easily pass through our bodies. When the rays hit the special x-ray film that Doris placed next to Molly's tooth, a picture of the inside of the tooth was made. This allowed Doris and Dr. Giles to see the cavity and infection there. In modern dental offices, x-ray photographs of teeth are rapidly replacing the traditional x-ray.

Ear Pain: Often, when we have a toothache, we feel pain in our ear on the same side as the bad tooth. For example,

Molly felt pain in her left ear as a result of the infection in her lower left tooth. The reason for this ear pain is that the nerve to the tooth runs by the ear. As Dr. Giles said, many people think they have an ear infection when they really have a tooth infection. Thinking that he or she has an ear problem, a person may go to a doctor that specializes in ear problems (called an ear, nose, and throat doctor). If no ear infection is found, the person will be sent to a dentist to see if a tooth is the cause of the pain.

Niki

I

Life is good. That's what Niki Maples (her given name was Nikesha, but everyone had called her Niki for as long as she could remember) thought as she was brushing her teeth at the sink in the tiny bathroom, in the tiny apartment where she lived with her parents and two younger brothers. Her attitude might have puzzled someone who didn't know Niki, or know that she always found something positive in any circumstance. Those who knew her knew that Niki was a force to be reckoned with. Yes, she lived in a neighborhood in which people thought twice about going out alone at night. Yes, her family didn't have much money. Yes, there were bad influences all around her. But Niki didn't let these things hinder her. She was determined to overcome all obstacles on the road to her success.

This evening, for example, she had completed almost three hours of required homework, and had even done some extra reading on her favorite subject, astronomy. Someday, hopefully, she would be an astronomer. She would study the universe and see and understand its beauty, even when she couldn't see beauty here on earth.

Now, as she was brushing her teeth before going to bed, she looked in the mirror, and even in the dim light of the bathroom she liked what she saw. She had a brown, flawless face with what people said were perfect features. Her boyfriend, Anthony, said the color of her skin was milk chocolate. He often called her Candy Bar. He thought she was beautiful and wasn't shy about telling her so. She enjoyed his compliments, but always reminded him that the true worth of a person included much more than how they looked. She was planning to meet him, as usual, in the morning, when they would walk together to school. Officially, they weren't going steady, but neither dated anyone else.

Niki spit the excess toothpaste into the sink and looked at her mouth. She especially liked her smile. As she moved her lips this way and that to examine her teeth, Niki remembered the years of braces required to produce the beautiful smile she had. She also appreciated the financial hardship the orthodontic work had placed on her mom and dad. She loved them immensely for sacrificing so much so that she could have a perfect smile.

She brushed her black wavy hair to remove any snarls that were in it. She had a habit of twirling her hair between two fingers when she concentrated her mind on a task, so by the end of the day she usually had several snarls that needed brushing out. As she continued her self-examination in the mirror, she might have asked herself: what was the end result of studying hard, eating right, exercising daily, and her careful grooming? The answer was that she had become an intelligent and healthy young woman of seventeen. That's right, a woman. Some seventeen-year-olds, including several of her friends, were still girls, but not Niki. She was so mature and responsible that anyone would say she was a woman, and deserved to be treated as one.

It was time to leave the bathroom, before one or both of her little brothers started banging on the door loud enough to disturb the neighbors. Niki left the bathroom, walked down the short hallway, and turned left into her bedroom. She felt fortunate that their apartment had a small spare room intended for a pantry that her parents converted into a bedroom for her. Niki closed the door and literally leaped onto the bed. With the door closed, her room became her castle. She felt safe and secure within those four walls. Spring wasn't quite here yet, and nights were still chilly, but her bed was comfy-cozy. As she pulled her quilted blanket over her, she was ready for sweet dreams. Yes, life is good.

II

"Yo...Niki...Niki." Niki turned a smiling face toward Anthony's familiar voice. He was slowly jogging toward her. At five feet eleven and two hundred ten pounds of solid muscle, he had the classic build of a fullback. In fact, he was the starting fullback on the Higgins High School football team. They were undefeated last season. Yet, as Niki watched him come toward her, she was impressed by the gracefulness he exhibited. If Anthony were a wild animal, he would be more a leopard than a buffalo. As he got closer, his jog became a walk.

"Want me to carry your books?"

"Of course, what else are men good for?" She laughed as she handed him her backpack.

"For more than just carrying books," he retorted. "What are you doing after school?"

"Going to my dentist. My jaw has been hurting lately, and I want him to check me out."

"Dentist...yuck."

"Don't say that. Dr. Lazarus is a great dentist. He placed a sealant on my teeth so that I never have cavities, and he referred me to my orthodontist, who did a wonderful job of straightening my teeth. You love my straight teeth, don't you?"

"Yeah."

"Are you going to come with me?"

"Yeah."

"Great. See you after school. And don't hang with your buddies; I have to see Dr. Lazarus at three, and I don't want to be late."

III

"Stop being such a fraidy-cat and come here."

Niki was at the entrance of the dental office of Dr. Lazarus.

"I told you, I don't like dentist's offices."

Anthony was standing on the sidewalk, and really didn't want to enter the office.

"He isn't going to touch you. It's me he's seeing."

Anthony slowly approached her, and they entered the office together, holding hands.

The reception room of Dr. Lazarus' office was lighted with ceiling lights and a few well-placed lamps that were not too bright. Soft music, typical of a dental office, was playing. Niki, with Anthony in hand, reported to the receptionist seated behind the glassed opening.

"Niki...er...Nikesha Maples here to see the doctor."

"Have a seat, Niki. Good to see you again. You're looking very well."

"Why, thank you. I'm glad to be here. Recently, I've been having pain in my mouth."

"Your mother called and gave us verbal permission to examine and treat you as necessary today." The receptionist looked down and appeared to be reviewing what must have been Niki's chart. She looked up at Niki and asked, "Have there been any changes in your health history since your last visit?"

"No."

"In that case, please sign your chart right here," the receptionist said, pointing to an area at the bottom of Niki's chart. She handed Niki the chart on a clipboard and a pen. Niki took the chart, signed it, and handed it back.

"Is the young man behind you also here to see the doctor?"

Anthony looked alarmed, and appeared ready to turn around and run out of the office.

"Heavens, no!" Niki said with a smile. "Although I'm sure he needs a good dental examination." She tugged on Anthony's arm, pulling him close to her. She then guided him toward a comfortable couch, sat him down, and sat next to him.

IV

The reception room was crowded with people, but it was only a few minutes before a dental assistant appeared and asked Niki to come with her. Anthony, still holding her hand as if he were permanently attached, followed closely behind. They went into a room that had an x-ray machine in it. The dental assistant, who introduced herself as Kelly, explained the need for an x-ray. Although Kelly was rather young, probably in her early twenties, she was obviously very competent.

"You're new here?" asked Niki. "I don't recall meeting you before."

"Yes," replied Kelly. "I've been here about three months, but I worked for Dr. Simon in Parkway City for almost two years prior to coming here. I really like it here. Dr. Lazarus is a wonderful person to work with."

Kelly continued, "I'm going to take what we call a panoramic x-ray, which will display your entire mouth on one x-ray film. That way, we'll be able to see all your teeth, including your wisdom teeth. Niki, please stand on this platform." She pointed to the area where she wanted Niki to position herself. "Young man, please step outside the room so that you will be shielded from the x-rays." Anthony reluctantly receded from the room, but stayed close by in the hall.

Kelly placed Niki's head into brackets that were part of a tall x-ray machine. Once she was satisfied with Niki's position and had put a special drape around her, Kelly told Niki to remain still while the machine rotated around her. Kelly walked out of the room holding a long curly cord with a button, which she pressed when she was completely outside the x-ray room.

The x-ray machine made a soft whirring noise and slowly revolved around Niki. When the machine stopped, Kelly returned and led Niki away from it. "Follow me, please."

Niki and Anthony followed Kelly down the hall and into a dental examining room. She was seated in a dental chair with all its associated equipment. If she let her mind wander, she could imagine it being in a rocket ship, taking her to visit the stars and planets she so loved. Anthony took up residence in the corner behind and farthest from the chair.

"Dr. Lazarus will be here in a moment. Nice meeting you, Niki."

"Same here."

Kelly left the room.

"Where are you hiding, Anthony?" Niki playfully chided him.

While Anthony was thinking of a clever reply, Dr. Lazarus entered the room.

"Hi, Niki," said Dr. Lazarus. "I see you brought another patient for me."

"No...I...I," Anthony stammered.

"Don't worry, young man, I'm only teasing."

"Better watch out, Dr. Lazarus, or Anthony will run out of here right over you."

"My goodness, I don't want to be on the receiving end of that. I've seen him run over linemen twice my size." This was the most humor that Niki had ever heard from Dr. Lazarus. He was usually quite serious and matter-of-fact.

"Okay, enough fun. Time for business," he said. But Niki could still see half a smile on Dr. Lazarus.

Dr. Lazarus put on gloves and proceeded to examine Niki's mouth. He used various tiny instruments and asked her questions such as, "Does this hurt?" or, "Have you been having pain here?" Niki replied no to these questions, although she did feel vague discomfort when he pressed on the back of her lower jaw.

"Everything looks fine in your mouth and on your x-ray, Niki, except that your wisdom teeth are starting to make their presence known."

Both Niki and Anthony looked concerned, and Dr. Lazarus noticed. "There's nothing unusual, Niki. Most people have wisdom teeth, and most people, at about your age, have to have them removed." He rolled his chair to the x-ray that was on a lighted view-box. "You can see the wisdom teeth here and here." Dr. Lazarus pointed to the four wisdom teeth visible on the panoramic x-ray. "It's important to remove them for various reasons, including preventing movement of other teeth. Since you've had orthodontic treatment, you don't want the wisdom teeth spoiling your smile. I could say more about wisdom teeth, but removing them is usually done by a specialist called an oral and maxillofacial surgeon. I recommend that you see Dr. Ralph Banner, who will discuss wisdom teeth in more detail and, I hope, take them out for you."

"Do I have to have my wisdom teeth taken out?"

"Have to...no. Should...yes. Your wisdom teeth are going to seriously bother you sooner or later. The best time to have them removed is before they cause problems. If they act up, extracting them can be more difficult. Basically, you are at the perfect age, and now is the perfect time to extract your wisdom teeth."

Niki didn't say anything—she was thinking—and Anthony couldn't have spoken if he had tried.

Dr. Lazarus pressed the intercom button and asked Kelly to bring Dr. Banner's business card. Kelly returned with one and handed it to Niki. "Here is Dr. Banner's name and telephone number," said Dr. Lazarus. "Consult with him to learn more about wisdom teeth, even if you decide not to have them taken out at this time. You'll like him. I went to dental school with him, and I know him very well."

Niki, who now appeared quite composed, looked directly at Dr. Lazarus. "Thank you, Doctor. I'll take your advice and see Dr. Banner. If he also recommends that my wisdom teeth be removed, I'll do it."

"Good decision, Niki. As I said, I'm sure you'll like Dr. Banner." Dr. Lazarus said goodbye and left the room.

Kelly assisted Niki from the dental chair and accompanied her and Anthony to the exit. "After you are finished with Dr. Banner," she said, "we will want to see you. He will let us know when he's finished with the treatment that you and he decide upon."

Niki and Anthony walked out of the office into the cool afternoon air. "What do you think? Do we see Dr. Banner and have my wisdom teeth out?"

Anthony was wiping sweat off his face with his sleeve. "What do you mean, we? I barely made it out alive today, and you want me to go to this oralmaxillowhatever with you? Are you crazy?"

Feeling playful, Niki pretended to be a deranged monster. "Yes, Anthony, I sure am." Using the scariest voice she could muster, she demanded, "Of course you're coming with me!" Niki giggled and started skipping down the sidewalk. Anthony, momentarily taken aback by her deranged monster impression, hesitated. Then he ran after her, before he lost her in the crowd at the crosswalk.

V

"How was your dental visit?" Mrs. Maples asked while scooping a pile of spinach and onions onto Niki's dinner plate.

"Fine, Mom. No problems other than I might need my wisdom teeth taken out. I have the name and number of a surgeon that Dr. Lazarus recommended. His name is Dr. Banner."

"If Dr. Lazarus recommended him, I'm sure he's excellent. Give me his telephone number and I'll call in the morning and make an appointment for you."

"Thanks, Mom, I'd appreciate that."

"Anything for my favorite daughter."

"Mom...I'm your *only* daughter."

VI

Niki and Anthony had taken a bus across town and were now walking the last few blocks to Dr. Banner's office. They were in a much more fashionable part of the city than the one they had left. The streets were lined with trees. Well cared for townhouses on both sides of the street were, in most cases, family dwellings. Others discreetly housed professional offices, and one of these offices was Dr. Banner's.

"Someday, I'm going to live in an area like this," Niki said to no one in particular.

"I know you will," Anthony replied.

Niki thought, This is one of the reasons I love him. He's my biggest fan.

"It should be just a little farther on the left," she said aloud.

They stopped at a gray stone townhouse. Three stone steps led to a small porch and a double-door entrance. A brass plate to the right of the doors read, in large letters: Ralph Banner, D.D.S., M.D. The second and third lines, in smaller font, read: Oral and Maxillofacial Surgery, and 51 West Street.

Anthony looked nervous, but Niki opened the door and went in followed by Anthony.

Classical music greeted them as they entered a tastefully decorated reception area. The chairs were soft and comfortable. There were paintings on the walls. Niki was impressed with what she saw. "My goodness, this is classy."

"Yeah," Anthony said.

Niki presented herself to one of the two receptionists. Her mother had given permission for treatment when she made the appointment. A chart had been created at that time. Niki and Anthony went to a small sofa and sat down. Niki relaxed, absorbed the atmosphere, and listened to the music. She knew the Mozart concerto that was playing, and hummed softly along with the music.

After a while, a nurse named Lucy, one of several nurses in the office, escorted her into an examining room and seated her in a dental chair. Once again, Anthony accompanied her and stood in an empty corner, even though a chair was available for him. Saying that Dr. Banner would shortly see Niki, Lucy left the room. Niki noted that an x-ray similar to the one Dr. Lazarus had taken was on a wall-mounted lighted viewing box.

A few minutes later, a rather young-looking, handsome African-American doctor appeared. Anthony was leaning in the corner and, for no reason that he was conscious of, came to attention, as if a commanding officer had entered the room. Niki fluffed her hair with a quick swipe of her hand, to ensure that she looked her best.

"Hi," Dr. Banner said, assuming they knew who he was. He walked directly to Anthony and shook his hand, taking Anthony's hand in both of his for a moment. As their eyes met, he said, "I know you. You're the fullback over at Higgins, right?"

"Yes…sir."

"I've seen you play. You're very good. Bet you get a scholarship to a major football school."

Puffing out his chest with pride, Anthony said, "I sure hope so."

Dr. Banner let go of Anthony's hand. He grabbed a stool, rolled it next to Niki, and sat down. "Dr. Lazarus sent me a note and the x-ray he took. The note says that you probably need your wisdom teeth extracted. Have you any thoughts regarding their removal?"

"Most likely, I'll follow your advice. If I should have them out, I'll do that."

"Dr. Lazarus said that you went to dental school with him," Anthony interrupted.

Dr. Banner looked at Anthony, smiled and replied, "That we did, young man."

"Anthony! Don't bother the doctor. He's got more important things to do," Niki scolded.

"Not at all. Yes, both Dr. Lazarus and I went to Meharry Dental School."

"Me-Harry?" Anthony repeated.

"Close enough. It's part of Meharry Medical College in Nashville, Tennessee. The interesting thing is both the medical and dental schools are mostly African-American, and have traditionally graduated a great percentage of the African-American dentists and physicians in the country."

Niki perked up at this news. She was intrigued by what Dr. Banner had said. "African-American schools?"

"Yes, Meharry was founded many years ago in order to train African-American dentists and physicians. Today, its dental and medical schools are filled with people of all races, although the professors are mostly African-American. It's one of the few institutions in the U.S. geared toward professionally educating us." He gave Niki and Anthony a conspiratorial look. "Another is Howard University in D.C., but Meharry has a special place in my heart."

"Wow!" Niki and Anthony said in unison, both feeling an indescribable pride at learning there were professional schools run by people of their race.

"I know how you're feeling. Today, people of color are well received by dental and medical schools throughout the country, but there was a time when that wasn't the case. If it weren't for Meharry, there would be far fewer African-American doctors."

Dr. Banner paused for a moment before continuing. "After Dr. Lazarus and I graduated from Meharry, I went to Miami and spent the next five years training in my specialty of oral and maxillofacial surgery while George...er...that is,

Dr. Lazarus, spent two years in the military, gaining great experience before he opened his practice of general dentistry here. When I decided to come to this city, I was very happy to learn that Dr. Lazarus was here. We often work together in treating patients. Now you know the whole story."

Both Niki and Anthony were still glowing about what they had been told. Such feelings were not unusual for Niki, who was well aware that African-Americans were successful in all walks of life such as business, politics, medicine, and astronomy, the field she hoped to enter one day. But they were unusual for Anthony. He was, except on the football field, a laid-back type of guy, unconcerned with issues such as race relations and job opportunities. The story that Dr. Banner had just related made him proud to be an African-American man. The two doctors he had met in the last few days were impressive. They had obviously worked hard to become what they were. He was proud of them, too. Maybe, if he worked hard enough, he could be a doctor, like Dr. Banner or Dr. Lazarus.

Dr. Banner felt that it was time to turn his attention to Niki's wisdom teeth. "Let's begin your evaluation, Niki. The first thing is to evaluate your x-ray." He looked at the x-ray on the wallboard. "This x-ray shows your jaws, teeth, and other structures, such as your sinuses, nose, and eyes. When you look at the wisdom teeth," he said, pointing to four teeth at the edges of the x-ray, "You get the impression that there is no room for them and that they are pushing on other teeth. They will never come in properly." Dr. Banner turned back to Niki. "By the way, the other name for wisdom teeth is third molars. They are called wisdom teeth because they, supposedly, erupt at the age of wisdom—that is, at about your age." Dr. Banner smiled and said, "A great many parents seem to disagree with that theory."

Niki craned her neck in order to get a better view of her teeth on the x-ray. She could see at a glance that Dr. Banner was correct. Her wisdom teeth were pushing on the teeth in front of them.

Dr. Banner continued, "We say that these teeth are impacted. This means that they are being blocked from properly erupting by other teeth and the jawbones." He took a laser pointer from his pocket and used it to indicate on the x-ray what he was talking about. "I recommend that they be removed, because, in my experience, they can cause damage such as infection with associated pain and swelling. I feel that they can also cause movement of other teeth by putting pressure on them that may undo some of the orthodontics you've obviously had. There are oral and maxillofacial surgeons who may disagree with my reasons for taking out wisdom teeth, but what I've told you has been my experience."

At this point, Niki knew she was going to have her wisdom teeth extracted. She wasn't going to risk having her other teeth move because of pressure from her wisdom teeth. She had spent too much time, and her parents had spent too much money getting her teeth where they were. "Why do we have them?" She had to know.

"My patients often ask that question, Niki. I'll answer by saying that we—that is, those who study such things—think that wisdom teeth will someday disappear. I don't know when that might happen, but it seems that it will. Mother Nature has already made our jaws smaller than they used to be. You can prove that by visiting a museum that has ancient human jawbones. You will notice that they are larger than the jaws we have. Those larger jaws could hold the thirty-two teeth humans have. Our modern jaws, being smaller, can usually hold only the twenty-eight teeth that erupt by about age twelve. The wisdom teeth that should erupt in the late teens—or as I said, the age of wisdom—don't have room to properly do so. Consequently, they become blocked or impacted. They may come in a little bit, but not enough. Unfortunately, Mother Nature wants them to continue trying to erupt, and that's when I've seen them move other teeth around." He stopped to give Niki and Anthony time to appreciate what he had said, then continued. "That's the bad news. The good news is that removing the wisdom teeth corrects the situation. All the potential problems go away, and the jaws become stronger as new bone forms in the extraction sites. Any questions?"

"No," both Niki and Anthony said.

"Good. Now let's talk about how we are going to take your wisdom teeth out."

Niki adjusted herself in the chair to be sure she would not miss anything Dr. Banner said.

Dr. Banner looked at her and asked, "Has Dr. Lazarus ever given you local anesthesia?"

"Yes, once, when he took out a loose baby tooth."

"Well, you can have your wisdom teeth removed the same way. I can give you local anesthesia in each wisdom tooth area. Then I can take the teeth out without pain. You would be awake, but you would only feel pressure and hear noise. You would not feel pain. Some patients opt for this method."

Dr. Banner paused for a moment before he continued his presentation. "Another option I can offer you is the use of intravenous heavy sedation."

Anthony, who was still in the corner, came a bit closer in order to better hear. "I put medicines in the intravenous fluid to sedate you. Putting you in outer space is what I call it. Most people have no memory of the extractions. That's just the way the medicines work."

Niki and Anthony smiled at Dr. Banner's description. Dr. Banner saw their smiles and said, "When I give you the medicines, you'll smile just as you're doing now." They both smiled again.

"When you don't care about anything, I'll give you local anesthesia. You won't know it. Also, the sedation, as I give it, is very safe. This is my most important requirement of the anesthesia. My next requirement is that you have a positive experience. I should be able to have you in and out of my office within forty-five minutes or so."

Niki and Anthony registered surprise at this last statement. They had thought it would take much more time to remove the wisdom teeth and recover enough to go home. "Wow! Only forty-five minutes?" Niki asked.

"Yes, Niki. The newer sedation medicines leave your body quickly once the surgery is over. You will walk out of the office a short time later."

Anthony looked at Dr. Banner, then at Niki. "It seems pretty easy, a lot easier than I thought. I'd have the sedation anesthesia."

Niki couldn't believe what she was hearing. Anthony, the world's biggest chicken in a dental office, was telling her that having her wisdom teeth out was easy. She was speechless, and could only smile at Anthony's advice.

Dr. Banner looked at him and also smiled. "You're right, Anthony, it is pretty easy. Most people have a good experience; ninety-nine percent, I'd say. Those few patients that have a less than ideal experience tell everyone about it and needlessly scare people. Fortunately, the people they scare usually have a good experience and realize that there was nothing to worry about."

"Nobody has frightened me...yet," said Niki. "And I *will* have the intravenous sedation anesthesia."

"That's great. As I said, you should have an easy time of it."

Anthony, now standing just behind Niki, placed a reassuring hand on her shoulder. Niki continued to be amazed by the apparent change in Anthony's well-known fear of dentistry. He not only seemed to have lost his fear, but he was becoming, unbelievably, a source of strength for her. Again, she could only look at him and smile.

Dr. Banner stated, "There is more I have to tell you regarding wisdom teeth extractions. There are nerves that run through both sides of your lower jaw. Each gives feeling to your lower lip and chin on one side."

Niki and Anthony were paying close attention.

"Because the nerves can be very near the roots of the wisdom teeth, it's possible to have a period of changed feeling or numbness of one or both sides of your lower lip and chin. It's rather rare, and on the few occasions I have seen it, feeling has returned to normal. There have been reports that the numbness can be permanent, but I've not seen that happen. Your teeth are not the type that typically causes this problem. If, while extracting your teeth, I see potential nerve injury, I may leave part of the tooth behind in order to minimize that possibility. The bottom line is that the situation is a function of anatomy that neither of us can change, and I'll do my best to see that your nerves are undamaged."

"I hope that doesn't happen to me."

"Very unlikely, Niki." Dr. Banner paused a moment. "You should also know that after you have your teeth out, you will need to take it easy for several days. Because of the sedation, it is required that you not drive yourself home. We will give the person that drives you home after the surgery both written and verbal instructions on how to take care of you postoperatively. The reason you don't get the postoperative instructions is that the medicines I give you will, most likely, cause you to forget about everything that happens here."

"Cool," said Anthony, almost causing Niki to fall out of her chair in disbelief.

"Do you have any questions, Niki?" said Dr. Banner.

"No, you've answered all I can think of." She turned toward Anthony and said mockingly: "Perhaps you have a question for Dr. Banner?"

Anthony lifted both hands in the air in a gesture that said further discussion was unnecessary.

"Well then," Dr. Banner continued, "if you have a question between now and the time you come to have your teeth extracted, please give me a call. I'll be happy to answer any question. If you would like, Lucy can set the date of your extraction appointment." Dr. Banner gestured to Lucy, who had reentered the room.

"I've just thought of a question."

"Go ahead."

"Will my insurance cover the surgery?"

"You have excellent insurance that will cover most of the surgery and the entire anesthesia," Lucy answered. "Come with me and I can make your appointment. As Dr. Banner said, you will need a ride to take you home after the surgery. Usually, one or both of your parents are the best choices for a ride. I'll give your mother a call and review your insurance coverage with her. Lastly, one of your parents will have to sign the permission for surgery form I'll give you. That's only because you're less than eighteen years old."

Niki and Anthony followed Lucy to the front desk where a receptionist made an appointment for Niki's surgery.

VII

Three weeks passed, and Niki and Anthony were once again at the door of Dr. Banner's office. Only this time, Jack and Phyllis Maples, Niki's parents, accompanied them.

"I hope he's all you say he is, Anthony," said Jack Maples.

"He is," replied Anthony as he opened the door for the others.

"You seem to like him," said Phyllis Maples.

"I do," answered Anthony. "He's the first dentist I haven't been afraid of. I'm even thinking of letting him take my wisdom teeth out."

At this revelation, Niki looked upward and rolled her eyes. "Now I've heard everything." She turned to Anthony. "You? You, Anthony Brown, are thinking of having your wisdom teeth removed?"

"Yes…if you have a good experience, I'm going to do it. I'm probably going to have to do it someday."

Niki, who felt she knew Anthony better than he knew himself, thought that having not eaten breakfast because of the scheduled surgery was making her hallucinate. He couldn't be saying what she thought he was saying.

The four of them took seats. They were the only people in the reception room.

"Nobody's here," Anthony noted.

"That's because this is Niki's reserved surgery time," said Lucy as she entered the room. "No one else is scheduled at this time. You must be Mr. and Mrs. Maples. I'm Lucy Graham, Dr. Banner's head nurse," she said, extending her hand. "I'll be assisting Dr. Banner with Niki's anesthesia. Kristy and Jim, our surgical assistants will also be helping with her wisdom teeth extractions. The doctor will be here in a few moments to review Niki's surgery with you."

"Nice to meet you, Lucy," said Jack Maples as he shook her hand. Phyllis Maples smiled and said hello.

At that moment, Dr. Banner entered the room and introductions were made. Dr. Banner reviewed the anticipated surgery with Niki, Anthony, and the Maples. He went over all that he had discussed on Niki's first visit. He talked about the surgery, anesthesia, postoperative care, and problems such as infection and nerve damage. He also confirmed that Niki had not eaten in the last several hours. He asked if anyone had any questions. The only question was Anthony's. He asked if Dr. Banner wouldn't mind sewing Niki's lips shut to keep her from talking for a while. Everyone laughed. Dr. Banner said that it was a common request from friends of his patients, but it would not be possible.

"It looks like Niki is ready to have her teeth out," said Lucy. "Please come with me, Niki."

Niki stood up from her chair. Her mom and dad kissed her and wished her good luck. Anthony also kissed her, but said that since Dr. Banner was taking out her teeth, he didn't need to wish her good luck. Dr. Banner was good luck enough.

Niki followed Lucy into the room where the surgery was to be done. It had the expected dental chair and a great deal of other equipment. Lucy had Niki sit in the dental chair. She then placed a blood pressure cuff on Niki's left arm above the elbow. Pressing a button on a machine caused the cuff to inflate and take her blood pressure. A small plastic clip was placed on a finger of her right hand. Lucy said the machine to which it was connected revealed the level of oxygen in her body. It also showed her pulse rate. Dr. Banner noted that the machines, called monitors, showed Niki to be in excellent health.

Lucy placed a bib over Niki's chest. A small plastic tube for oxygen was gently positioned just under her nose, allowing the oxygen to flow into her nose. Dr. Banner, standing on Niki's left, asked her to hold out her left arm, elbow down, and place it on what he called an arm-board that was attached to the chair. There was a handle to hold. Dr. Banner said he was going to start an intravenous line in order to give Niki fluids and medicines during the anesthesia. Niki was a bit nervous, but she was pleasantly surprised at how quick and painless the procedure was.

"I'm going to begin giving you some medicines that will relax you. In a moment, you will be sedated enough for me to take out your teeth."

A feeling of well-being came over Niki. She felt very tired, and wanted to fall asleep. She was vaguely aware of the events around her, but not the least bit concerned about them. Suddenly, Lucy was telling her that the surgery was over, and all was well.

Niki had had no awareness of time passing. It seemed that she had just sat in the dental chair. Her head began to clear and her eyes began to focus. "How long did it take?"

"About thirty-five minutes," replied Dr. Banner.

Niki couldn't believe that thirty-five minutes had passed. "I don't remember a thing."

"That's the way it usually goes, Niki. You're biting on cotton, and I don't want you to talk too much," he said.

Niki couldn't feel the cotton she was biting on. In fact, she couldn't feel anything in her mouth. Lucy saw her touching her chin and explained, "Your mouth is numb from the local anesthesia that Dr. Banner gave you."

"Oh," was all Niki could reply.

Dr. Banner and Lucy allowed Niki to remain seated for the next few minutes while she became more alert.

"I think I was asleep."

"You were in a natural sleep for a few minutes, but you were mostly sedated. In fact, you opened and closed your mouth when we asked," said Dr. Banner.

"Amazing! I don't remember a thing."

Dr. Banner responded, "As I said on your consultation visit, that's the way the medicines work. That's why we gave all the home care instructions to your parents in the reception room as I was taking your teeth out. You wouldn't remember them."

Niki nodded that she understood.

Lucy said, "Niki, please sit up and turn toward me."

Niki did as instructed.

"How do you feel?" asked Lucy.

"A little woozy, but fine."

"Good. Now stand up."

Niki stood up as directed. She was steady and able to accompany Lucy into the reception area. Anthony was there, but her parents were not. She turned to Lucy with an expression that said, "Where are my mom and dad?"

Lucy understood Niki's expression. "They're getting the car. Anthony and I are going to walk you to it."

Anthony, relieved to see how well Niki was doing, came to her. Lucy and he, one on each side of Niki, walked her to the car.

VIII

During the drive home, Niki recovered from the anesthesia, but was still somewhat relaxed. Things had gone well. She knew that it would be a while before she was normal again, but if she followed Dr. Banner's instructions, full recovery was just a matter of time.

As she rested against the backseat of the car next to Anthony, who was holding her hand, a feeling of euphoria came over her. The last few weeks had been a learning experience. She had learned all about wisdom teeth and their removal. She was pleased that she had gone through with the surgery. As Dr. Banner had said, it was a lot easier than she had expected.

She had also learned about African-American dental and medical schools and the doctors they produced. What wonderful institutions! But the most pleasant surprise was Anthony. In the beginning he was so frightened, but—as fantastic as it seemed—he had lost his fear of dentistry, and became a source of encouragement for her. Niki turned to him. "Thanks for all you've done. I might not have gone through all this without your help."

"What're you saying, girl? There was nothing to worry about with Dr. Banner in charge. And guess what?"

"What?"

"While you were having your teeth pulled, I made an appointment with Dr. Banner's receptionist to have my wisdom teeth checked out. And if they need to come out, I'm going to have them taken out just like you did. I'll go to sleep for a few minutes, and when I wake up, they'll be gone."

"Really, Anthony?"

"Sure. I know I can do it." He continued talking softly, soothingly, no louder than if he were humming a tune. He didn't want to disturb Niki too much. "You know...I really like Dr. Banner. I wouldn't mind being like him. I might even go to Meharry."

Niki closed her eyes and thought: Dr. Banner not only took out my teeth. He helped Anthony believe in himself and to know that he is much more than a good football player. I thank Dr. Banner for doing that, and, most of all, I'm so proud of Anthony. These were her last thoughts before the rhythmic motion of the car lulled her to sleep.

Definitions: Niki

Erupted and Impacted Teeth: All teeth are made of a crown, which we see and use for chewing, and a root, which holds them in the jawbone. Erupted teeth have their whole crown visible. Impacted teeth are either partly erupted (i.e., only part of the crown is showing), or not erupted at all (i.e., none of the crown is showing). They are almost always third molars, which are also called wisdom teeth. Dr. Banner has found that impacted teeth can cause problems such as infection and pushing of other teeth. Because of these problems, many people must have them taken out.

Panoramic X-ray: A panoramic x-ray shows the entire upper and lower jaws and teeth. The x-ray itself is rectangular in shape and about the size of a loaf of bread. The advantage of this type of x-ray is that it shows all the teeth on one film, thereby minimizing radiation to the patient. Impacted teeth and their relationship to other teeth are easily seen.

Anesthesia: Three types of anesthesia are used in oral and maxillofacial surgery:

Local Anesthesia: This is the classic injection given to put teeth to sleep. Only the mouth is affected. The patient is aware of what is happening, but feels no pain from the procedures performed.

Intravenous Sedation: An intravenous fluid line is started. Sedative-type medicines, such as tranquillizers and narcotics, are given through the intravenous line. The patient is technically awake and can usually respond to verbal commands, although he/she usually will not remember anything. This is a very safe anesthesia with few side effects.

General Anesthesia: The patient is completely asleep. Oral and maxillofacial surgeons use this type of anesthesia, but there is a trend toward the sedative types of anesthesia. Management of general anesthesia is more difficult than with sedative anesthesia, and there are more potential side effects. Note: Nitrous oxide, laughing gas, is often used as a supplement to these anesthesias.

Safety Equipment: The oral and maxillofacial surgeon's office will have, at a minimum, the following safety equipment:

Oxygen: The most important drug in the oral and maxillofacial surgeon's office. It is used in virtually all emergency situations.

Pulse Oximeter: Monitors the amount of oxygen in the blood through a small clip placed on a finger. This instrument also monitors the pulse rate.

Blood Pressure Monitor: Takes the patient's blood pressure at a preset interval, such as every five minutes.

EKG Monitor: Records the patient's heart rhythm.

Automatic External Defibrillator (AED): Used to shock the heart in the event of cardiac arrest.

Emergency Drugs and Equipment: The office will stock drugs and equipment to handle any type of emergency, from allergic reactions to cardiac arrest.

Oral and Maxillofacial Surgeon: The modern oral and maxillofacial surgeon spends four to six years in specialized training. He/she is trained in all aspects of the medical and surgical care of a patient. The oral and maxillofacial surgeon concentrates on surgery of the head and neck, including facial cosmetic and reconstructive surgery, repair of facial fractures, removal of cysts and tumors, placement of dental implants, and extraction of teeth. The oral and maxillofacial surgeon also trains in administering anesthesia, and may spend up to a year in the general anesthesia department of a hospital. Today, many oral and maxillofacial surgeons leave training with both a dental degree and a medical degree.

Complications of Extracting Wisdom Teeth: When one has wisdom teeth removed, he/she can expect to limit activities for several days, be on a soft diet, and take medicines such as antibiotics, pain medicines and anti-swelling medicines.

Most often, the patient does very well and heals without complications. Rarely, a patient will develop an infection that will need to be vigorously treated. Another complication that all patients have to be advised of is postoperative numbness of the lower lip, chin, or tongue. These conditions are the result of the proximity of the lower wisdom teeth to the nerves that run through the jaw. The oral and maxillofacial surgeon will do his/her best to avoid these nerves, including leaving part of the tooth behind, but in some cases avoidance is impossible. If a patient develops this complication, the numbness will usually disappear with time. In a small percentage of cases, the numbness can be permanent.

Insurance: Many insurance companies cover the extraction of wisdom teeth. Coverage will vary depending upon the particular insurance company. Prior to surgery, it is wise to ask the insurance company for a predetermination of benefits in order to know how much the insurance company will pay for the anticipated surgery. The patient will be responsible for whatever balance is left after the insurance payment.

Andrew

I

"Let me tell you about Andrew." The room was very quiet. You could have heard a butterfly fluttering.

"Go ahead," he said.

"Well, I feel that Andrew, since birth, has had trouble moving his tongue" she continued.

"In what way?"

"For one thing, he can't stick it out very much. The front seems to be stuck."

He showed no reaction, as if such tongue problems were not unusual. "I see...anything else?"

"Yes. He has trouble saying certain words."

"Can you give me an example?"

"He has difficulty with words that start with T. I took him to a speech therapist, who recommended that I make an appointment with you, Dr. Rogers." While Mrs. Patton was speaking, Andrew sat quietly beside to her. They were facing Dr. Tom Rogers, an oral and maxillofacial surgeon, who was sitting behind his desk. Dr. Rogers had noted that Andrew, who was only seven years old, was well behaved in the dental office.

Mrs. Patton, who appeared to be about forty years old, was dressed in a dark, conservative designer suit. Her auburn hair was fashionably done.

Andrew wore a shirt and tie with dark corduroy pants. His tousled hair was chestnut brown and his eyes were expressive. He was solidly built for seven years old.

"Are his lips dry most of the time?"

"Now that you mention it, Andrew's lips are usually dry and often chapped. I'm forever putting lip balm on them."

"Thank you, Mrs. Patton. I think it's time for me to have a look at Andrew."

Dr. Rogers was middle-aged, short, balding and a little overweight. One would say that he was not physically impressive. He pressed a button, and an assistant, who was professionally dressed in a brightly colored scrub uniform, soon appeared in the consultation room. Her youthful appearance contrasted sharply with that of Dr. Rogers.

"Sara, will you take Andrew and his mom to the examining room, so I can get a good look at Andrew's tongue?"

"Certainly, doctor."

II

Andrew and Mrs. Patton followed Sara to the examining room. Andrew was instantly attracted to Sara. She had an air of caring about her. Sara assisted Andrew into a rather modern-looking dental chair, while Mrs. Patton sat herself in a nearby chair. She then prepared Andrew for Dr. Roger's examination by placing a drape over his chest and repositioning the dental chair a bit. When finished with these tasks, Sara excused herself and left the room.

Within a few moments, Dr. Rogers entered the room. When last seen, he had been wearing a suit and tie. He now sported the classic surgical scrub suit so often seen on TV medical programs. Dr. Roger's potbelly was even more evident.

"Hello again," he said, nodding to Mrs. Patton as he breezed by her on his way to the examining chair. He pushed a few buttons and the chair responded, putting Andrew in a semi-reclining position. Dr. Rogers put on a mask and rubber gloves, and adjusted the overhead light so it would shine directly into Andrew's mouth when he opened it.

Taking a small dental mirror in his left hand and positioning himself on Andrew's right side, Dr. Rogers asked, "Would you please open your mouth wide?"

Andrew did as directed.

Dr. Rogers moved his mirror around Andrew's mouth, pushing out on both cheeks, down on the back of his tongue, under the sides of his tongue and, finally, lifting up under the front of his tongue.

"Andrew, open your mouth wide and stick your tongue out as far as you can."

Andrew did as directed.

"Stick your tongue to the top of your mouth, just behind your front teeth."

Andrew complied.

"Now, lick you lips."

Andrew tried to lick his lips, but was unable to do so.

"Mrs. Patton, come stand by me while I have Andrew move his tongue."

Mrs. Patton walked over and stood next to Dr. Rogers (her three-inch height advantage became quite evident), as he had Andrew repeat the tongue movements.

"See how the middle of the tongue doesn't move as much as the sides do when he moves his tongue forward or up. It reminded someone of the bow that Cupid uses to shoot his arrows, so we call it a Cupid's Bow."

Surprised at the appearance of Andrew's tongue, Mrs. Patton nodded and said, "I've never noticed that before."

"Can I shoot arrows out of my mouth?" Andrew said, laughing at what he thought was a very funny joke.

Mrs. Patton smiled, and Dr. Rogers muttered an acknowledgement and continued. "Also, Andrew can't lick his lips. Licking our lips is something we should be able to do. Otherwise, our lips get dry and cracked."

"My goodness." The smile was gone.

Dr. Rogers turned to Mrs. Patton. "The reason Andrew has difficulty saying certain words is that he has a condition that we call ankyloglossia, or what is commonly called tongue-tie. He can't move his tongue to the proper place for pronouncing some sounds. Usually, it's a problem that is easily corrected."

Mrs. Patton appeared visibly relieved at these last words.

Dr. Rogers turned back toward Andrew.

"Lift your tongue up again, as far as you can, please, Andrew."

Andrew did so. This was getting to be fun. How often can you stick out your tongue at people without being scolded?

"Andrew can't protrude or lift his tongue as he should, because a tiny film of tissue under the tongue is holding it back. This tissue is called a frenum. There's one under the tongue and others under the upper and lower lips near the front teeth. I find that patients call them 'strings'. We all have them, but sometimes they are bigger than they should be."

With a small piece of gauze in one hand, Dr. Rogers lifted Andrew's tongue to expose the frenum. "It's the frenum that creates the Cupid's Bow I mentioned." His other hand held an instrument that he used to point out the thin triangular-shaped tissue to Mrs. Patton. "If I trim the frenum, the tongue won't be restricted from its normal movement, and the problems he has with pronouncing certain words and licking his lips should be able to be resolved. I should mention that the Cupid's Bow normally remains for a time, until the tongue internally corrects itself. Postoperative tongue exercises will help."

Mrs. Patton felt goose bumps as she asked, "Does it hurt?"

"No. In fact, in the realm of surgery that I do, correcting a tongue-tie is considered a minor procedure."

"Oh, good."

"While I can administer a sedative-type anesthesia, usually this procedure is done with local anesthesia."

"Just Novocaine®!" exclaimed Mrs. Patton.

"Yes, if the patient cooperates. And from what I've seen of Andrew, he would be very cooperative."

Andrew smiled.

"I know he will be. He's a very good boy, and very good in a doctor's office. Do I sound like a mother?"

"You sound just fine. Using local anesthesia simplifies things, as I can move his tongue as needed without worrying about blocking his breathing. And the procedure only takes about fifteen minutes from start to finish."

"That's fast," said Andrew.

Both Mrs. Patton and Dr. Rogers were surprised at Andrew's unexpected response.

"Yes, Andrew, it is, but in your case that's all it would take for the correction."

Dr. Rogers positioned himself to face both Mrs. Patton and Andrew, but continued looking at Andrew, "And, when I'm finished, you'll only have a few dissolvable stitches to indicate you had anything done. I almost forgot; you will also be minus one excessive frenum."

"Will he still have control of his tongue?"

"Even better than before, Mrs. Patton. That's the point of the surgery. I want Andrew to have better use of his tongue, and not the restrictions that he has now."

"How about afterward? Will he have much pain? What difficulties will he have?"

"Good questions. Let me tell you about the postoperative phase of Andrew's treatment."

Mrs. Patton paid close attention.

Dr. Rogers continued, "After the surgery, Andrew will need to limit his activity for a few days. He will have to eat a soft diet, and may need over-the-counter acetaminophen appropriate for his age. In addition, I will recommend a cream that can be applied to the surgical site in order to minimize discomfort. You don't have to remember what I'm telling you today, as detailed verbal and written instructions will be given on the day of the procedure. Finally, since we will be using local anesthesia, Andrew can eat prior to the appointment, but I recommend a light meal."

"That's wonderful," said a smiling Mrs. Patton. "How do I arrange to have the correction done, and will it be covered by my insurance?"

"Sara, will review your insurance coverage with you and, and my receptionist will schedule a surgical appointment for Andrew." Turning to Andrew, he asked, "Do you have any questions? Have you understood what I've said?"

"Yes," Andrew almost shouted.

"Very good." Dr. Rogers pressed a button, and Sara appeared within a few moments.

"Sara, Andrew requires a correction of a tongue-tie. Please review Mrs. Patton's insurance with her, and see that an appointment is made for Andrew's surgery."

"Yes, Doctor." Turning to Mrs. Patton, "As I was aware of Andrew's problem and its correction, I've already reviewed your insurance plan, Mrs. Patton. The surgery is fully covered except for a small deductible that is due at the time of service. If you like, we can make an appointment for the correction of Andrew's tongue-tie."

"I'd like to do that."

"Okay, then follow me." Sara led Mrs. Patton and Andrew to the front office, where an appointment was made.

III

"Good morning, Professor and Mrs. Patton. Good morning, Andrew."

"Good morning, Dr. Rogers," all three replied in unison.

Addressing Professor Patton—whose great height, beard, and tweed clothing made him a poster person for college professors—Dr. Rogers asked, "Do you have any questions, Professor? You weren't here for the consultation visit, so if you have any, let me know."

"I don't think so. My wife explained everything." He so towered over Dr. Rogers that he was looking nearly straight down as he spoke. Had Dr. Rogers not been there, the professor would have been looking directly at his Hush Puppies®.

To Andrew, he said, "How about you, Andrew? Any last-minute questions?"

"No."

"Okay then, Sara will take you to the procedure room."

Professor and Mrs. Patton, and Andrew followed Sara down the hall to the designated room. When Dr. Rogers arrived, Andrew was sitting in the dental chair with a drape over his chest. The Pattons were seated in chairs against one wall. Sara, who had already dressed for the procedure, helped Dr. Rogers into his surgical gown.

He said to the Pattons, "I assume you will stay in the room while I do the surgery."

Both Pattons nodded and remained seated.

"Fine."

"I'm going to assist Dr. Rogers, Andrew," said Sara. "First, I'm going to place a numbing cream under your tongue, to make the local anesthesia that Dr. Rogers will give you less uncomfortable." Sara placed a small cotton swab under Andrew's tongue. "Its mint flavor will make your tongue feel tingly."

Andrew noticed that his tongue did, indeed, begin to tingle.

Dr. Rogers said, "You may notice a small pinch when I give you the local anesthesia." Dr. Rogers gave Andrew the local anesthesia. He felt, as Dr. Rogers had said, only a small pinch. It really didn't hurt at all.

"Your tongue may feel big because of the local anesthesia, but it isn't. It's the same size as it always was. In a few minutes we'll get started, and nothing will hurt you."

Dr. Rogers and Sara stepped away and sat in their chairs while waiting for the anesthesia to work. After a few minutes, both returned to Andrew, one on each side. The surgery took about ten minutes. Andrew felt nothing other than a little pulling of his tongue.

"All done," said Dr. Rogers for all to hear.

In truth, Andrew had been quite nervous about the surgery. He knew that he would benefit from it, though, and it was important that he be helpful. Now, he realized that there had been nothing to worry about.

"Professor and Mrs. Patton, please come and stand next to me," said Dr. Rogers.

The Pattons did as directed.

"Andrew, please stick your tongue out as far as you can."

Even though his tongue was numb, Andrew was able to do as requested. He even wiggled it from side to side to show off.

"Wow!" said Mrs. Patton, as she saw that Andrew could now place his tongue well out of his mouth. Professor Patton's smile showed that he was also pleased.

"You'll notice that Andrew is able to move his tongue much farther out of his mouth than he could before. He now has normal tongue movement. You will also notice that there is still a Cupid's Bow at the front. As I said to you on the consultation visit, Mrs. Patton, that has to do with the internal structure of the tongue. It should lessen or go away completely as time goes by."

"I'm so pleased, Doctor," said Mrs. Patton.

Dr. Rogers thought he noticed tears welling up in Mrs. Patton's eyes. "And," he continued, "with proper speech therapy Andrew's tongue should not hinder his speech."

Andrew sat quietly while the adults around him seemed impressed with the fixing of his tongue.

Sara went over the postoperative instructions with the Pattons. They were mostly commonsense rules that Andrew would easily be able to follow. The stitches under his tongue were dissolvable and would take care of themselves. When she was finished with the verbal instructions, she handed Mrs. Patton a printed sheet of the instructions, saying that if there were questions, they should call the office. She turned to Andrew. "How are you doing?"

"Good." He had a happy face.

Sara lowered the dental chair so that Andrew would have no difficulty getting out of it, and helped him do so. Professor and Mrs. Patton hugged Andrew tightly and told him what a good boy he had been.

Andrew smiled. He was clearly happy.

Sara and Dr. Rogers escorted Andrew and his parents out of the office, waving to them as they left.

Sara spoke first. "Isn't he the most wonderful special needs child you've ever seen?"

"I must say, I was really impressed with him," replied Dr. Rogers.

"Autistic children are amazing. They can be absolutely brilliant in certain areas of knowledge. I once met a fifteen-year-old who might have been the world's leading authority on U. S. presidents. She knew every tiny detail about them," said Sara.

"We don't know what causes the condition," said Dr. Rogers, "but I hope we find out soon, because approximately one out of one hundred and sixty five children is born autistic. In the past, many of these people were stigmatized by society and were, on occasion, placed in institutions. Today, we know that with love and care, they live happy, productive lives. I'm sure Andrew will do wonderfully."

"I hope so...for Andrew's sake," said Sara. She turned and walked back into the office.

Dr. Rogers thought he noticed tears welling up in Sara's eyes.

Definitions: Andrew

Ankyloglossia: A condition in which the attachment in the front of the tongue is larger than normal. It's positioned in the front of the under surface of the tongue. The result is that movement of the tongue is limited in all directions. Speech is often hindered. Dry lips result, as the tongue can't moisten them.

Tongue Anatomy: The tongue is made of five muscles functioning as one entity. They are the genioglossus muscle, the hyoglossus muscle, the styloglossus muscle, the longitudinalis verticalis muscle, and the transverses muscle. The tongue is attached on the inside of the lower jaw in the front and on both sides.

Mandible: The horseshoe-shaped lower jawbone that holds the lower teeth. The upper jawbone is called the maxilla.

Lingual Frenum: The technical name for the excess attachment under the front of the tongue. It is part of the tongue muscles.

Cupid's Bow: Because of the large lingual frenum, the front center of the tongue doesn't move as far forward as the front sides do. This constriction in the center gives the front of the tongue, when looked at from above, a shape similar to the numeral three. Someone thought it looked like the bow that Cupid uses to shoot his arrows and called it a Cupid's Bow.

Dorsum and Ventral Surface of the Tongue: The top of the tongue is called the dorsum or dorsal surface. It is covered with various papillae. The underside of the tongue is called the ventral surface.

Papillae (singular: papilla): These are the small bumps on the dorsum (top) of the tongue. They come in four varieties: filiform, fungiform, vallate, and foliate. They help sense objects such as food, and contain taste buds that help us taste food.

Nerves: The tongue has various nerves that help it do its job. The names of these nerves are the lingual nerve, the vagus nerve, the glossopharyngeal nerve, the internal laryngeal nerve, and the hypoglossal nerve.

Speech: Placing the tongue in certain parts of the mouth is required for proper pronunciation of words. If tongue movement is restricted, speech is often impaired.

Speech Therapist: A person who treats speech problems. Often, the speech therapist is the one who discovers that tongue movement is restricted and refers the patient to the oral and maxillofacial surgeon.

Chad
Part 1

I

Mom and Dad, his girlfriend Mary, his dog Blade, and mountain biking in his beloved New Hampshire mountains were the things that were most important in Chad Northrop's life. While biking was fifth on his list, it wasn't far behind the others. Riding in the woods, with Blade running after him, Chad could feel the burn in his legs as he raced up a mountain trail—could smell the clean air, the flowers, and the other fragrances of the outdoors. Enveloped in quiet and solitude, with the only noise being his labored breathing and the sounds of his mountain bike fighting the terrain, Chad felt joy in every fiber of his slender, hard, six feet two body. He knew this natural high was better than any artificial high that drugs could provide.

When Chad wasn't biking for fun and when he could fit it into his schedule, he was biking competitively. He was quite successful as a competitive racer, and was known for his aggressive riding throughout northern New England. Those who followed the sport said he was one of the best. Otherwise, Chad was a mild-mannered young man of twenty-two years.

In the snow season he was a ski instructor at Wild Mountain, a family-run ski facility fifteen miles from home. Wild Mountain was somewhat off the beaten path. Great skiing, but minimal amenities kept away skiers from Boston and New York, which was just fine with the local folk. On any given day during the season, the slopes were filled with skiers from neighboring towns, and school busses brought children to the slopes after school. The rosy cheeks of the children were like beacons lighting up the trails. Chad especially enjoyed teaching children. They followed his instructions as best they could and had no fear of doing what he asked. Ski instructing provided the income that, during the remainder of the year, allowed him to mountain bike many of the trails he skied. That, and living with his parents.

Yes, he was still living, as they say, "at home". Chad told himself that it was a way of saving money. He had been looking at a ten-acre piece of land near the top of one of the smaller mountains in the area. It had been used for grazing by a sheep and goat farmer. The farmer had retired and the land was for sale. The views from it were spectacular. He wanted to buy the parcel and build a home. Then he'd ask Mary to marry him. He wondered if she would choose Mary Northrop or Mary O'Brien-Northrop. It really didn't matter what name she chose, he was just curious. At the moment, Chad was a few thousand dollars shy of the parcel's asking price.

Until Chad actually bought the land and began building a house, ski instructing was his only real work. Once the ski season was over, he had a great deal of free time, much of which he devoted to mountain biking. When he wasn't racing competitively, Chad and several of his biking buddies would arise before dawn. After a hearty breakfast, they would put their bikes and equipment in one of their pickups and drive several miles to one of their favorite biking trails. They'd begin the trek as the very top as the sun began revealing its first rays of light. The trails were mostly in shadow at that time of the morning. Add moisture from dew and gravelly rocks and you had the makings of an interesting ride. His fair skin, as white as a baby's bottom his friends said, would take on a ruddy glow from the exercise. Chad and his friends went full bore until about ten o'clock. That's when the supply packs that were found in almost every part of a mountain bike were opened. A campfire was made, and in no time the smell of coffee permeated the woods.

After one such exhilarating ride, Chad was leaning back against a tree stump and holding a steaming cup of coffee with both hands. To no one in particular, he said, "It doesn't get any better than this."

"Ah, the Chad 'It doesn't get any better than this' speech," Alex stated. "Quicker than usual." The comment made Douggie and Mo laugh.

Chad grabbed a handful of dirt from the base of the stump and slung it at Alex, missing by a wide margin. More laughter from Douggie and Mo.

"What do you mean, the 'Chad speech'?"

Mo chimed in before Alex could answer. "Because every time we're at our first break—no matter if it's biking, skiing, fishing, whatever—you always say, 'It doesn't get any better than this.' No offense, we love to hear you say it."

Chad smiled. "I didn't realize I say that but it's true, especially with biking. I feel the burn in my legs, and I become more alive. The burn becomes a necessary part of me." Chad was referring to the burning feeling in a mountain biker's thighs that occurs from a buildup of lactic acid in the muscles. It tells an athlete that the workout has been hard.

"Ditto," said Douggie. "I know what you mean."

Mo nodded in agreement.

"I didn't know I was going to start a touchy-feely thing," said Alex. "Let's hit the trails before we begin to recite poetry. Sorry I brought it up in the first place."

"'But, soft! What light through yonder window breaks?'" Mo began, but started to chuckle and couldn't continue his Romeo and Juliet© balcony-scene impression.

Alex lifted himself off the ground and began to brush his pants of debris. This action signaled that he was ready to ride and the group should follow.

"Gee, Alex," Mo said, "just because Douggie and I are gay and enjoy Shakespeare doesn't mean that straight macho guys like you can't enjoy it."

Douggie and Mo had been a couple since their mid teen years. Douggie, six feet and a hundred and ninety pounds, had the muscles of a collegiate wrestler, which, indeed, he had been while in college. Mo, on the other hand, was slim and wiry, with muscles like spring steel. Chad often said that Mo was potentially the best biker of their group. He just didn't dedicate himself as Chad did.

Douggie and Mo were "cultured." That's what their friends said. Aficionados of classical music, art, poetry, and great literature, they were invited to every party or intellectual gathering. No matter the event, Douggie and Mo attracted crowds wanting to hear their views on almost any subject.

Both Alex and Chad, lifelong friends of Douggie and Mo, accepted them—as virtually everyone did—without reservation. There were no uncomfortable moments. Douggie and Mo were able to discuss any aspect of their gay lifestyle with Chad and Alex. On more than one occasion, Mo or Douggie sought advice from their two close straight friends that would have been difficult to seek from their gay ones. The four of them might as well have been brothers. That's why Mo could tease Alex about his machismo attitudes. He knew no offense would be taken.

Alex's action stirred the group. One by one the other three stood, cleaned up the area, carefully put out the fire, and returned to their bikes. Alex was Greek-American. His most distinguishing feature was his dark, intelligent eyes. They were coals set in a face masked by a full beard. Women said he was strikingly handsome. Alex was also the unspoken leader of the group. At twenty-six he was the oldest and, having been married for several years, the most settled. He organized their riding excursions, and usually the group followed his lead on the trek.

"It's always difficult to get back on a bike after relaxing or eating," Douggie said.

"Come on, Douggie," said Alex. "I thought you were a hard-core rider. You should be able to jump on your wheels and take off like a rabbit running from a hound."

"I'm like Douggie," Chad responded. "A good nap would be appreciated. Even with the coffee, I bet it's going to take me ten minutes before I'm back to full speed."

The group had taken their break close to where the trek divided into several trails. The trails snaked up Pine Ridge Mountain, frequently crossing one another until arriving at the ridge for which the mountain was named. The view from the ridge made the trip worth it. Each man picked a trail and began the upward climb.

Chad's trail was narrow and gravelly, with large rocks on either side. Add the trees and brush, and he had the impression that he was alone on the mountain. This was the feeling he craved. He knew that just a few feet away were other riders on their trails, but as far as he was concerned, they were a million miles away. In these conditions, his mind went into a Zen-like state and he rode the trails as if by instinct.

"Watch it!"

Chad awoke from his reverie only an instant before he collided with Mo. Mo was attempting to cross Chad's trail from the left when his front tire hit the back tire of Chad's bike. Mo was able to keep his balance, but Chad's bike spun counterclockwise, ejecting Chad in the process. He was thrown face-first into a large rock on the side of the trail and then bounced onto the gravel.

"Ohhh," Chad very slowly began to roll over and sit up. His first reaction was to see if he had broken any bones. He could taste blood, so he knew that his mouth was bleeding. He was glad he was wearing his helmet—it had probably saved him from more serious injuries. He took the helmet off and placed it by his side.

"Easy, buddy." Mo had immediately jumped off his bike and ran to attend to Chad. "Where does it hurt?"

"Everywhere...but my mouth hurts the most."

Mo took off his backpack and began rummaging through it. He found a hand towel and placed it in Chad's mouth in an attempt to stop the bleeding. "Bite on this," he said.

Chad complied.

After a while, Mo said, "Now, open your mouth so I can take a peek." Chad did, and Mo looked in. "It looks like you lost a tooth."

"Oh, no! Which one?"

"One of the upper front ones...the left front one, I think."

"Just my luck!" Chad momentarily took the towel from his mouth. He noted that several areas of his body were throbbing. "How's my bike?"

Mo looked over the bike where it lay across the trail. "Not much damage that I can see. It'll need a little fixing."

"There's some good news, at least. Look in my mouth again. See if you can tell what happened to my tooth. I don't think I swallowed it." He removed the towel—his mouth didn't seem to be bleeding any longer.

"Okay, open wide." Moving his head this way and that in order to get a good view, Mo examined Chad's mouth closely.

"I see a dark red area where the tooth is missing, but I don't see a tooth. I think it got knocked out."

"Look by the rock," said Chad, "maybe its there."

Mo moved toward the large rock to which Chad referred. He noted a bloody area on the rock where Chad's face had hit it.

Douggie and Alex appeared, having come back to search for their friends when Mo and Chad didn't make it to the ridge. "What's going on?" asked Alex.

Douggie, seeing that Chad had obviously been injured, jumped off his bike and ran to help him.

"We collided," Mo said, not turning as he continued his search about the rock. "I'm looking for the tooth Chad lost."

While Douggie did what he could for Chad, Alex immediately went to help Mo. After a few moments, Mo called out, "I found it!"

Alex went to where Mo was and knelt beside him. Mo pointed to the tooth. They both stared at the perfect, undamaged tooth, which was lying on a leaf in a pile of gravel.

"Pick it up, Mo," said Alex

Mo seemed paralyzed.

Watching them, Chad hollered, "Well, someone pick it up! I'm going to need it put back in my mouth."

In response, Alex took a handkerchief from his back pocket and checked to see that it was relatively clean. Using only the handkerchief, he picked up the tooth and wrapped the handkerchief around it.

Chad's head was slowly clearing, and he was beginning to think about his circumstances. He knew that it was important to treat the tooth gently if it was to be put back in his mouth—that is, if it was even possible to do that.

"Does anyone know how I should protect the tooth until I get to a dentist?"

"I read somewhere, in a health magazine I think, that you should put the tooth in milk or under your tongue until you can get treatment," Douggie replied. "Does anyone have any milk left? I don't have any."

"I threw mine away after our break," replied Alex.

"So did I," Mo chimed.

"I don't have any either, so I guess it'll have to be option two," Chad said. Grinning through his pain, he joked, "Who wants hold it under their tongue for me?"

His three friends appeared close to vomiting.

Still grinning, he said, "I guess I'm going to have to do it. Bring me the tooth."

Alex walked to Chad, who was now standing. He gently unfolded his handkerchief, exposing the white tooth with what appeared to be a red film around its root. Carefully taking the tooth from the handkerchief, Chad placed it under his tongue. Holding his tooth with his tongue caused his speech to be garbled, but his friends understood him when he said, "Take me to a dentist...now!"

II

Chad found that he was able to walk, with a little help from Douggie, who was the strongest of the group. Mo and Alex walked the four bikes down the mountain. Alex had used his cell phone to call 911, and an ambulance was waiting for them on the approach road when they got to the base of the mountain. All the while, Chad kept the tooth under his tongue.

He was admitted to the Walker Regional Hospital Emergency Room and was immediately evaluated by a petite nurse in a blue surgical scrub suit, who took his vital signs. As time is critical when a tooth is knocked out and needs to be reimplanted, Dr. Maxwell Barre, an oral and maxillofacial surgeon, had been called regarding Chad's injury. Dr. Barre had immediately cleared his office schedule and gone to the emergency room. He was at the hospital when the ambulance carrying Chad arrived.

Once it was determined by the admitting nurse that Chad's injuries were not urgent—that is, requiring immediate life-saving treatment—and that he had no head or spinal cord injuries, he was quickly turned over to Dr. Barre's care. A technician had taken the knocked out (technically, avulsed) tooth from Chad and placed it into a container of liquid specially formulated to preserve such teeth. The container had been placed on a sterile surgical tray containing all the instruments Dr. Barre might need to replace the tooth in Chad's mouth. Chad, lying on a stretcher, was wheeled to the waiting Dr. Barre.

"Hi, Chad. I'm Dr. Barre, an oral and maxillofacial surgeon."

Chad looked up from the stretcher and saw a distinguished-looking man, about forty-five years old, with black hair that was graying at the temples and a thin mustache. Although Dr. Barre was of medium height, his narrow waist and broad shoulders made him appear taller. He obviously spent some of his free time working out in a gym. He had on a blue scrub suit that was, apparently, the hospital's typical uniform.

Dr. Barre continued, "You lost what we call your upper left central incisor tooth. In other words, your left front tooth. Fortunately, your friends retrieved it, and I'm going to put your tooth back in your mouth."

Chad smiled to himself, remembering how squeamish his friends had been with the tooth.

"Because the success rate of a reimplanted tooth depends, to a great extent, upon how soon it is replaced after being lost, I'm going to begin treatment immediately. I've examined your tooth, and it appears to be an ideal candidate for reimplantation. I'm doing the procedure without x-rays, but we'll get them when you come to my office for your postoperative visit."

Dr. Barre had an assistant and a surgical setup ready to go in one of the surgery areas of the emergency room. Outside the room, Chad noted the hurried activity that was the hallmark of busy emergency rooms everywhere. The overhead speakers were constantly playing announcements, asking some doctor to call some telephone extension. But inside the tiny surgical room, pulling the two flimsy drapes closed created a surprisingly quiet retreat from the rest of the facility.

Without delay, Dr. Barre applied a numbing cream to Chad's upper mouth, followed by an injection of local anesthesia. Within five minutes Chad's face and upper jaw were asleep. He had the feeling that his face had gotten fat. Dr. Barre said this feeling was normal with local anesthesia.

Dr. Barre cleaned Chad's face and upper jaw with an antiseptic solution. Next, he took the tooth from the liquid in which it had been stored. In what seemed to be one swift motion, he inserted the tooth back into its socket. Dr. Barre's eyes told Chad that all was well. The next few minutes were spent applying wires—Dr. Barre used the term "arch bar"—to Chad's upper jaw. The arch bar would hold the reimplanted tooth in place while it healed.

As Dr. Barre took off his gloves and mask, he said, "All done."

Chad and his three biking buddies left the hospital a short while later. In his pocket were prescriptions for an antibiotic and a pain medicine, a sheet of home care instructions, and an appointment card noting that he was to see Dr. Barre at his office the following week.

"Guess who's going to try to get Mary to feel bad for him and be his little nurse?"

"I hear you, Mo—but I'd feel bad for you, and I'd be your nurse," said Douggie.

"Ooh, Douggie."

"Ooh, Mo."

They couldn't contain their laughter, as they hugged each other and pretended they were Chad and Mary. They continued all the way to the parking lot.

III

The emergency room staff had called Mary, Chad's person to call in the event of an emergency. Tires squealing, she pulled into the parking lot just as the boys were leaving the hospital. Completely disregarding the parking lines and leaving the car door open, she ran to the four friends. "What happened?" Anxiety filled her sky-blue eyes.

Chad was beginning to bruise from his injuries. In spite of the pain medicine he had been given, he was in considerable pain, but he didn't want Mary to worry. "Just a little accident. I'm okay"

She wrapped her arms around Chad and held him. Tears were streaming from her eyes.

"Are you sure?"

"Yes, of course. Don't worry."

"What do you mean, don't worry? When the person you love gets injured, you worry."

Seeing an opening to lighten the mood, Chad countered, "Finally admitting you love me, huh? I'll have to get injured more often." He gave a big swollen smile.

Realizing that, in spite of his appearance, Chad wasn't seriously injured, Mary released him and took a step back. She had classic Irish beauty, with fair skin, curly red hair, and those sky-blue eyes. "You know I love you!" she said with feigned indignation. "If I didn't, why would I want you to be the father of my children? Uh...that is...our children," she quickly corrected.

"Are you proposing? And how many children are you planning on?" Chad retorted.

Mary gave him a sly grin. "Yes. And three...at least."

Listening to Mary and Chad, Alex could only say to Mo and Douggie, "It looks like our future is going to be filled with baptisms and birthdays."

"Poor Blade," Mo said, taking Alex's lead. "Kids will be climbing all over him."

Douggie chimed in, "Blade will love it. He'll be their doggy daddy."

"Boys," Mary announced, seeing her opportunity, "I'm sorry to break up this stimulating conversation, but Chad and I have to go. He needs rest in order to get well; and I need him well, so we can make wedding plans."

"Uh...Mary?"

"Yes, Chad?"

"Will it be Northrop or O'Brien-Northrop?"

Mary replied impishly, "You'll just have to wait and see." She took Chad's hand and led him to her car. As they walked away, Chad turned to his buddies and winked.

IV

Six weeks later, it was Chad's third visit to Dr. Barre's office, and he was hoping it would be his last.

Dr. Barre returned to the treatment room holding an x-ray in his hand. "The x-ray of your tooth looks good. At the moment, I don't see any problems. How does it feel?"

"Fine."

Standing beside Chad and shining his light in Chad's mouth, Dr. Barre examined the injured area. He pressed and prodded, and said things such as "Uh-huh" and "Umm." Chad was getting a little nervous. He hoped there were no complications.

Finally, Dr. Barre took a step back and looked at Chad. "You appear to have healed well. It's time to remove the arch bar that's holding the tooth in place. I'll give you local anesthesia and remove it. It'll take no more than five minutes."

"Do I have to get local anesthesia?"

"No, but it's much more comfortable with it."

Chad agreed to the procedure. Dr. Barre kept his word. After Chad's mouth was asleep, it took less than five minutes for the removal of the arch bar. When he was done, Dr. Barre said that Chad was finished with his treatment; and would, most likely, never have a problem with the tooth. But Dr. Barre continued, "Chad, several complications can occur in a reimplanted tooth. It can get abscessed, for example, and require root canal treatment."

"Root canal treatment," Chad repeated. "I've heard bad things about root canals."

"They're all lies, Chad. Root canal treatment is usually easily done and painless."

Chad felt he could believe what Dr. Barre said.

Dr. Barre continued, "Other things can happen that can cause you to lose the tooth. I'm referring to what are called internal and external resorption. These are conditions in which the tooth dissolves away, from either the inside out or the outside in. In these cases, the tooth is often removed. Internal and external resorption may occur years from now."

"After all I've been through, it would be a shame to lose the tooth," Chad responded.

"That's the bad news. The good news is that if you were to lose the tooth, you could have it replaced with an implant."

"An implant?"

"Yes. An implant is inserted into your jawbone to replace the root of the tooth. Once it heals, a crown is attached to the implant, making it a complete tooth. If you need one, I'll discuss it in much more detail at that time."

Chad was lost in thought for a few moments. "Well, Dr. Barre, I think that knocking out my tooth was the best thing that has happened to me."

"What do you mean, Chad?"

"Well, when Mary saw me walking out of the hospital after you treated me, she realized how much she loved me. And I knew how much I loved her. We began making wedding plans right there in the parking lot. So, in a manner of speaking, you're responsible for us getting married next month." Chad took an envelope out of his pocket and handed it to Dr. Barre, who looked puzzled. "It's an invitation to our wedding. Mary and I want you and your wife to come." (In casual conversation during one of Chad's postoperative visits, Dr. Barre had spoken of his wife.)

"I'll check with my wife. If we don't have other commitments, we'll be there." Dr. Barre placed the invitation in his jacket pocket and left the room. As an oral and maxillofacial surgeon, Dr. Barre helped people. Often the people he helped were in pain, or lying on a stretcher in the emergency room, or in some other predicament that needed his immediate attention. He felt good when he helped them. Today was different. Today, he felt wonderful.

Chad
Part 2

I

Chad seldom thought about his biking accident and the subsequent reimplanting of his tooth. Three years had brought great changes to his life. Mary and he had married and they had finally saved and borrowed enough money to purchase the parcel of land that Chad had been eying. The home he had built, largely with his own hands, would eventually need additions to accommodate his family, but at the moment, it was more than adequate for their needs. The views of the valley were nothing less than magnificent.

Mary and he had been blessed with twin boys about a year ago. Their names were Alex and Douglas. Their next boy—an event they did not doubt would occur—would be named Stephen, which was Mo's given name. Even Mo couldn't remember when he got his nickname; he had been called Mo for as long as he could remember. Needless to say, he had heard many times that it was because he would fit right in with Larry and Curly™. The twins were now in the crawling stage and kept their parents busy. The boys, Mary, and Chad were lucky to have very special uncles in Douggie, Mo, and Alex. On many occasions, one of them babysat to give Mary and Chad some needed relaxation time.

Another bit of good fortune came to Chad when the State of New Hampshire designated a large mountainous area at the edge of their village as a new state park. Because of his reputation as an outdoor person, Chad was able to secure the position of head park ranger. For Chad, this was the perfect job. He oversaw the daily management of the park and mostly worked outdoors in the clean mountain air. He was given full time use of a four-wheel drive Jeep®, but it surprised no one that Chad's trips around the park were often made on his trusty mountain bike.

Since it was early summer, and the park roads and trails were finally drying after the late spring rains, today Chad decided to do his rounds on his mountain bike. The last time he had used his bike for this task was just after Thanksgiving, before the snow had made biking too difficult. He occasionally rode his stationary bike during the winter, but it didn't provide the same invigorating feeling that riding outdoors did. However, riding the stationary bike prevented the saddle sores he might otherwise develop from his first outdoor ride of the season.

Chad left the cabin that held his office and took the road that veered to the right, leading to the higher terrain. There was about a half mile of straightaway with tall pines on either side before the road began to ascend and turn from macadam to hard dirt. Chad decided to ride at full speed for half of the straightaway and then coast to the dirt. As he began to coast, he stood on the pedals and lifted off the seat. The wind whipped around his face and helmet. Just as he took in a breath while his lips were somewhat apart, he felt a sharp pain in the front of his upper jaw.

"Whoa!" he yelped, and the unexpected pain almost caused him to fall off his bike. He immediately stopped, his legs on the road straddling his bike. Placing his finger under his upper lip, where the pain was, he felt a bump above the front upper tooth he had knocked out a few years earlier. Chad knew that something was wrong with the tooth. He turned his bike around and slowly rode back to his office. He would call Dr. Barre today.

II

Chad was given an appointment for later that same afternoon. Sasha, Dr. Barre's dental assistant, took an x-ray of the previously injured tooth, and Chad was seated in an examining room. Shortly after, Dr. Barre walked into the room with the x-ray on a holder.

"How are Mary and the twins?"

"There're all very well, but the boys are keeping us busy."

Dr. Barre placed the x-ray on a view-box and reviewed it. He turned to Chad. "I'm afraid there's a problem with the tooth that I put back in your jaw." Chad gave him a questioning look, but said nothing. "If you recall, on your last visit here, I mentioned that certain problems could occur with your tooth. It seems that one of them has occurred."

"What are you saying, Doctor?"

"I'm saying that your tooth is undergoing what we call external resorption. That is, the outside of the root is dissolving. You can think of it as a form of infection. Unfortunately, the tooth will need to be extracted."

"There's no way to save it?" There was concern in Chad's voice.

"I'm afraid not. On your last visit, I wrote in your chart that I mentioned the subject of implants. Do you remember?"

"Yes, I recall you said something about them."

"Well, now its time for me to really tell you about them." Chad was all ears as Dr. Barre continued. "To summarize, Chad, implants are pieces of titanium metal that are inserted in the jaw to replace the tooth root. Once the implant heals, a crown—patients usually call them caps—is attached to it. And presto! You have a new tooth. It's almost like having your original tooth."

"Wow!" was the first thing that came to Chad's mind.

Dr. Barre explained, "Implants can be placed in the socket of a tooth that has just been extracted, or we can wait a few months and insert the implant."

"How do you decide?"

"If I extract a tooth and can clean out all traces of infection, I'll place the implant at that time. If I can't clean out the extraction socket properly, I'll wait and place the implant at another time. In the upper jaw, it usually takes about six months for the implant to heal—what we dentists call osteointegration."

"Sounds as if it's different for the lower jaw," Chad commented.

"Yes, in the lower jaw it usually takes about three months."

Chad thought about what Dr. Barre had said. He then asked, "What do I do in the meantime? Do I go without a front tooth?"

"Don't worry, Chad. I'll be working with your general dentist, who will make a temporary replacement until the permanent one can be completed," Dr. Barre replied. "On your chart, you filled in Dr. Adams in Ellis Falls as your regular dentist. Is he?"

"Yes, I guess so. Although the last time I saw him was two years ago."

"That's okay. We'll make an appointment for you to see him, and I'll review your case with him. In the meantime, I'm going to put you on an antibiotic to clear up the infection you have and a pain medicine for discomfort."

"Oh, good," Chad said, relieved to hear about the medicines. He was hoping they would prevent the pain from coming back.

"So, Chad, Sasha will make two appointments for you. One with Dr. Adams and, after you see him, one with me to place the implant in your upper jaw.

"It sounds easy enough." Chad was trying to be macho, although he was a bit nervous about the procedure.

Having seen many patients over the years, Dr. Barre was aware of Chad's anxiety. "Let me tell you what to expect during the next visit." Dr. Barre reviewed with Chad all aspects of the placement of a dental implant. At the conclusion of Dr. Barre's presentation, Chad was visibly relieved.

"As I said, it sounds easy enough."

III

A few weeks later, Chad was once again sitting in Dr. Barre's dental chair. As Dr. Barre walked in, he said, "Hi, Chad. Dr. Adams tells me you and he have arranged for his part of your implant."

"Yes, I saw Dr. Adams and we discussed the implant. He's going to make me a temporary replacement tooth while the implant heals. He'll put a permanent replacement on the implant in about six months, when the implant is fully healed. I'm going to see him for placement of the temporary tooth after I leave here."

"Great," said Dr. Barre. "I know you were concerned that you might have to walk around with a tooth missing."

"You said it," replied Chad. "As I deal with the public, I'd be self-conscious if I had to greet people with a big hole in my smile where a tooth used to be."

"I understand, Chad, and it looks like everything is going to work out well. In a moment, Sasha will take you to the surgery room, and we'll get started."

IV

An hour later, Chad was on his way to see Dr. Adams. He had a new dental implant where his upper left front tooth had been. Shortly, he'd have a temporary tooth replacement, and his smile would be unchanged. As he drove the Jeep® to Dr. Adams' office, Chad reflected on what might be referred to as The Adventures of Chad's Front Tooth. It had started innocently enough with a bike ride with his friends. At the time of his accident, he hadn't been sure that he would ever end up with a good result in his mouth. As it turned out, he'd had two good results—first, when he had his own tooth reimplanted; and second, when he had the dental implant placed in his mouth.

Two other wonderful things had happened as indirect results of the accident. He and Mary had married, and they had their two children. Dr. Barre proved to be not only an excellent surgeon, but a friend as well. Dr. Barre and his wife had attended their wedding and, occasionally, each had been invited to the other's home for dinner. Dr. Barre had made the whole process easy for Chad.

Chad thought about the changes in his life. It seemed that just a short time ago, he had been a boy with minimal responsibilities and with a boy's intense enjoyment of life. Today, he was a man with a wife, children, a home, and a job, and all the responsibilities that come with those things. And, thankfully, he still had a boy's intense enjoyment of life.

He shifted gears, pressed the accelerator pedal, and felt the kiss of the wind through his hair as the Jeep® picked up speed.

Definitions: Chad

Traumatic Tooth Avulsion: Loss of one or more teeth from its/their sockets. If the socket retains its shape (morphology) and the tooth is not broken (fractured), it can be replaced (reimplanted) in its socket. The success of the reimplantation depends upon several factors, such as the time from loss of the tooth to replacement, the age of the patient, the condition of the tooth, and the preservation of the tooth in the interim. It is ideal to replace the tooth within twenty minutes of injury. Both the tooth and the socket must be intact.

Resorption: External/Internal Resorption is the loss or dissolution of tooth root structure. Traumatically injured teeth that have been reimplanted may have one of two types of resorption occur to them.

External Resorption: This is the loss of tooth structure from the outside to the center of the tooth.

Internal Resorption: This is the loss of tooth structure from the inside to the outside of the tooth.

Root Canal Treatment: A tooth is hard on the outside and soft on the inside. The soft inside, called the pulp, is in the space called the root canal. It contains, among other things, the nerve of the tooth. When a tooth is infected (abscessed), it is the pulp that is abscessed. With root canal treatment, the pulp is removed and replaced with a special root canal filling.

Dental Implants: Dental implants are used to replace missing teeth. Most dental implants are root-shaped objects made of titanium metal. The human body does not perceive them as foreign, and accepts them. As a result, bone grows tightly around the implants and makes them very solid. This process is called osteointegration. Once an implant has osteointegrated, a crown can be attached to it, thereby completely replacing the lost tooth.

Jon

I

Jon Yamata had many interests, and his family genealogy was one of them. His mother had been born in Liverpool, England, and had come to America as a teenager. His father, although born in America, could trace his Japanese ancestors back many generations to samurai and the feudal period of Japan. Jon loved to read about that feudal Japanese period. The tales of great battles and fighting for honor were exciting. His mother said that she had had similar feelings when, as a girl, she read about King Arthur, the Knights of the Round Table and, most of all, Queen Guinevere.

Jon, lying in his bed prior to falling asleep, often imagined himself back in feudal Japan. In his imagination, he vanquished his enemies with his sword, with his bow and arrows, and with his superior strength and knowledge of the martial arts. His sacred sword, forged by a long-dead master of sword making, was ever at his side and ready for action. Being an expert horseman, he would ride at top speed while shooting an arrow in the dead center of his target. And he could easily defeat several men at the same time with his lightning kicks and punches gracefully coordinated with blocks, jumps, and back flips.

In the real world, Jon maintained a relationship to his Japanese culture via his study of karate, and his goal was to someday be a Grand Master of the art. At seventeen years old he was already a black belt and beginning his climb up the ranks to the coveted red and white belt of Grand Master. It would take many years of extreme dedication, and, most likely, he would never obtain Grand Master status, but he was willing to strive for it.

The martial arts in various forms have been around since ancient times. Peasant farmers, perhaps in Okinawa, China, or India, who had to defend themselves against well-armed and armored cavalry, invented the self-defense arts. As they were not allowed to have weapons, they used their work tools as weapons and developed fighting techniques for them. Today, people see the modern versions of these weapon-tools and think them exotic. They don't realize that hundreds of years ago a weapon that is now used to subdue an opponent was used to plant beans.

Those farmers had to develop punches and kicks that penetrated armor or made it useless. With a punch, they were able to knock their armored opponents off their horses and deal with the enemy from a more favorable position. As these fighting techniques were developed in different areas of the Eastern world, variations of technique resulted. Consequently, there are many styles of the martial arts. Karate, kung fu, tae kwon do, and jujitsu are examples.

Jon chose karate (literally translated as empty hand) after a family visit to Japan, when he was ten years old. During that visit the family attended a demonstration at the dojo (training hall) of a famous karate sensei (teacher). He was amazed to see candles blown out with the force of the wind from a punch (seiken) and the tops of bottles broken off with the hand formed like a knife (shuto) without cutting the hand.

He was particularly impressed with the breaking of wooden boards, bricks and cinder blocks, and the day he broke his first brick is one he will always remember.

While Jon had many interests, he had one great dislike. Jon disliked the way he looked. The reason for his dissatisfaction was his large lower jaw, which protruded well beyond where it should have. When he closed his mouth, his lower teeth were in front of his upper teeth. This was different from his friends who, Jon noted, had upper teeth that were in front of their lower teeth. And his chin...what could he say about it, other than that it was a monstrosity?

Jon felt that people were always looking at his jaw and chin. His own experience was that if someone had anything unusual, such as one eye that looked to the side or a large hairy mole on their nose, he could not help staring at the defect. No matter how hard he tried to prevent it, his eyes would keep coming back to what he was trying to avoid. He thought that the same thing occurred with his chin—that is, people just couldn't avoid staring at it. Jon's chin was to him, as Cyrano de Bergerac's© nose was to Cyrano.

Jon was intelligent, an excellent athlete, and a fairly good conversationalist. He should have been brimming with confidence. But the truth was that his large lower jaw made him feel self-conscious. It drained any confidence he should have had.

Another consequence of his large lower jaw involved eating. First of all, his jaw made loud noises when he chewed. People heard the noises across the table. Chewing gum was impossible, as his jaw tired and became painful in a short period of time. Even talking, after a while, caused his jaw to hurt.

Recently, Jon learned of possible good news. A television documentary showed a corrective procedure for deformed jaws. The patient in the documentary had a problem similar to his. The procedure was performed by a team of oral and maxillofacial surgeons at a medical school in Texas. At the end of the program, people with similar problems were encouraged to seek an oral and maxillofacial surgeon in their area. He wrote down the information presented on the screen. After seeing the documentary, Jon felt a sense of hope. There might be a solution to the cause of his unhappiness, and he was going take action.

II

"Hi, Dad."

"Hello, son," said Jon's father, as he came into the house. Mike Yamata was a seventh grade mathematics teacher at Grayson Middle School, two blocks from their home. He felt fortunate that he could walk or bike to work. Southern California weather helped with this endeavor. When he biked, he got plenty of attention, as riding his bike and carrying his oversized briefcase necessitated a balancing act worthy of a high-wire expert.

Emily Yamata was a social worker for the town, and wouldn't be home for another hour or so. The Yamata home was a typical Spanish-style common in the suburb in which they lived. It was tastefully furnished and accented with English and oriental antiques.

"What's up?" said Mr. Yamata. It was unusual for Jon to be home when he arrived from work. In fact, Jon was always the last of the family to come home. After school hours were busy for him. He was on the school track and swim teams, and practices usually kept him at school until at least six o'clock. And two evenings during the week he went directly to the dojo for his karate class. Mr. Yamata knew that there was a good reason for Jon to be home at this time.

"I know what I want for my graduation gift!" Jon excitedly exclaimed.

"Isn't it a little early to be thinking of a graduation gift? It's only January," said Mr. Yamata, smiling. "What if you don't graduate?"

His dad knew, of course, that there was no doubt of Jon's graduating. Very few honor students didn't graduate. But one of the nice things about their relationship was that they often kidded and played practical jokes on each other.

Today, Jon wasn't in a joking mood. "I'd like it as soon as possible."

"Jon, this sounds serious." Mr. Yamata replied.

"It is."

"Tell me what you need." Mr. Yamata was anxious to know what Jon felt he needed immediately.

"Don't get nervous, Dad. It's good, not bad. Something that will improve me."

Mr. Yamata was visibly relieved. "What is it, son? Don't keep me waiting."

"I want corrective surgery of my jaw. I saw it on TV and researched it on the Internet. It's what I need."

"Corrective surgery! What kind of corrective surgery?"

"To reposition my lower jaw backward, so I can eat better."

"And look better?" said Mr. Yamata, knowing the answer.

"Yes, Dad...and look better."

Mr. Yamata was silent, wondering what to say next. He knew Jon was sensitive about his appearance.

"Tell me about the surgery."

"It's done by a doctor called an oral and maxillofacial surgeon. In my case, I think that some of the bone of my jaw would be removed, and the jaw pushed back to a more normal position. I really want this, Dad! I even have the name of a surgeon that can do the surgery."

Mike Yamata knew his son wasn't prone to snap decisions. If he wanted corrective jaw surgery, Jon would have researched the issue thoroughly before deciding to go through with it. If this surgery was what his son wanted, he would help as much as he could. "Okay, Jon. If Mom agrees, we'll make an appointment with the doctor."

III

Emily Yamata was very supportive of her son's decision to have corrective surgery. She wanted to end Jon's embarrassment about his jaw. She made an appointment with Dr. Jillian Penski. One sunny but cool afternoon two weeks later, she, her husband, and Jon were in a consultation room with Dr. Penski. Jon was seated in a dental chair, his parents were seated on a bench against one wall, and Dr. Penski was leaning against a counter next to Jon.

Dr. Penski, a petite, middle-aged woman not much more than five feet tall, began with, "I understand Jon wants an evaluation regarding possible surgery on his lower jaw."

The three Yamatas nodded.

"As I look at you, Jon," she continued, "I feel that you are a likely candidate for jaw surgery. We call such surgery orthognathic surgery. It refers to surgery of the jawbones. Your problem appears to be your overdeveloped lower jaw, called the mandible. Your upper jaw, the maxilla, may be part of the problem. I won't know that until I study your x-rays, photos, and plaster tooth models that we'll take today. That's all we'll do today." Turning her attention to Mr. and Mrs. Yamata, she said, "Once I come up with a diagnosis and treatment plan, I'll have you return, and we'll review options for correcting Jon's problem."

"Very good," said Mrs. Yamata.

"Any questions?" asked Dr. Penski.

Mr. Yamata responded, "Not at this time, but we'll probably have a few at the next visit."

"Okay, then." Looking at Mr. and Mrs. Yamata, she said, "I'll ask you to return to the reception room while my assistants take the necessary x-rays and other records on Jon. If you like, you can make the appointment for your next visit."

The Yamatas, pleased with this initial interview with Dr. Penski, stood, said goodbye and left the consultation room. They made another appointment to see her.

IV

Arriving for their second appointment with Dr. Penski, the Yamata family was anxious to hear what she had to say about Jon's problem. They were escorted to the same consultation room as the previous visit. Jon was seated in the dental chair, and his parents were seated on the bench against the wall.

The room contained more than it had on their last visit. There were x-ray photos on the wall. On the counter that Dr. Penski seemed to enjoy leaning against were plaster models of Jon's teeth. These were in a metal brace that held them in the same position as in Jon's mouth. Also on the counter was a computer and x-ray photos, with tracing paper over them.

"Welcome back," said Dr. Penski, walking into the room. She had on a white doctor's smock, with her name stitched in blue on the left breast pocket. The collar of her pink blouse folded over the collar of the smock. She had the professional appearance expected of a prominent doctor. She proceeded to the counter and leaned against it. Being a no-nonsense person and a busy doctor, Dr. Penski began the session immediately.

"I've had a chance to review your case," she said, looking at Jon, "and I can help you." She turned her attention to the x-rays. "Here are the x-ray photos of Jon's present condition." Several x-rays were placed on viewing boards; others were on the counter, covered with tracing paper with various lines, angles, and measurements.

She paused to give the Yamatas time to evaluate the x-rays.

"You can see that Jon's lower jaw," she stated, pointing to an x-ray, "is too far forward and angling downward. This situation is abnormal. The lower front teeth should be behind the upper front teeth."

Jon nodded, indicating that he knew the teeth should be as Dr. Penski said.

Continuing after a short pause, "We oral and maxillofacial surgeons say you have a prognathic mandible with a wide gonial angle. The only other problem is that your chin is too large. Fortunately, Jon's upper jaw, called the maxilla, is in a normal position. That makes any corrective procedure less involved."

This last statement brought small but noticeable smiles to Jon and his parents.

"What I mean," Dr. Penski continued, "is that my study indicates that only your lower jaw needs attention. In many cases, perhaps most cases, the maxilla has to be corrected along with the mandible."

Their smiles increased.

"The most astounding thing is that you don't need orthodontic treatment prior to surgery. This is an extremely rare situation. Consider yourself fortunate."

By now, the Yamatas were positively beaming.

Dr. Penski continued her review of Jon's problem and its correction. She reviewed the x-rays on the wall holders, and those on the counter covered with the tracing paper and markings. She showed the Yamatas the plaster models of Jon's teeth before and after the anticipated corrective surgery. The computer was used to illustrate what Dr. Penski called predictive surgery. Jon's profile appeared on the monitor as it presently looked, and how it would look after surgery. Before and after photos of patients she had previously treated were also shown. Many of these patients had more deformity than Jon. Seeing such patients and the good results they had encouraged him regarding his anticipated surgery.

"The bottom line," Dr. Penski said to all, "is that Jon needs his mandible set back and his chin reshaped. Before I go into detail about the surgery, are there any questions? At the last visit, you said you'd have some."

The Yamatas were in what can only be described as awe at what Dr. Penski had presented. They reserved their questions for later and asked Dr. Penski to continue.

Changing gears, Dr. Penski asked the family if they needed a break and, perhaps, coffee or tea. Only Mike Yamata accepted her offer and asked for coffee. Everyone moved a bit and relaxed. Dr. Penski excused herself and left the room.

After a few minutes, during which time Mr. Yamata's coffee was brought to him, Dr. Penski returned. She looked at Jon and said, "I will use what is called a sagittal split™ procedure to correct your problem. It allows me to do all my work from inside your mouth. There are no outside scars."

Jon nodded, and Mr. and Mrs. Yamata paid close attention.

"The horseshoe-shaped lower jaw is surgically cut on both sides in the back. When that's done, the front part that contains the teeth can be placed in its ideal position and held there—initially, with a combination of bone plates and orthodontic elastics. The elastics keep your teeth together. Patients often refer to this situation as wiring the jaws together. After approximately a week, the elastics are removed. The bones will soon heal in their new position. I'll reshape your chin by removing a horizontal wedge of bone from it. It will be smaller and shapelier. That surgery is also done from the inside of your mouth."

Dr. Penski paused to judge the Yamata's reaction to what she had said. By their expressions, they seemed to be accepting her presentation well. "All of this is done in the hospital. You would be admitted the day of surgery and stay for a few days afterward. A week prior to admission, I will do a physical examination on you. You'll need minimal blood tests done at the hospital laboratory. If your surgery were more involved, I would arrange for auto-transfusions. Auto-transfusions are the taking of your blood before surgery and giving it back to you at the time of surgery. You won't need them. Basically, your preadmission is quite easy. After you return home, you'll visit me once a week for six to eight weeks, so I can make sure that all is well. You should be able to return to school in less than two weeks after the surgery, although athletics will be prohibited—or at least restricted—until you heal. The only other thing I must mention is that the surgery involves the major nerves of the lower jaw. There is often a period of numbness of the lower lip or chin. In my experience, the numbness has always disappeared."

Looking at the trio, Dr. Penski asked again, "Now do you have questions?"

The Yamatas looked at each other, and then Emily Yamata looked at Dr. Penski. "Forgive me, but Mike and I have been wondering—you're not very large physically and the surgery seems strenuous. How can you do such surgery?"

"Technique, my dear Mrs. Yamata...technique."

V

The Yamatas had finished dinner and were sitting in their living room among the English and oriental antiques. Mrs. Yamata was the first to speak. "Jon, how do you feel about undergoing the surgery that Dr. Penski described?"

"I can do it. I want to do it. I will do whatever it takes." Jon had a determined expression. His mind was made up. The surgery would change his life for the better.

He had no fear of the surgery. His study of karate was more than physical. Mentally, it prepared him for any eventuality in life. Most people think that the martial arts consist only of punches and kicks. They are not aware that the true study of the martial arts is mostly a mental discipline. The martial artist trains his or her mind in order to gain inner strength. For Jon, karate had prepared him for his surgery.

VI

"Welcome to Mainland Memorial Hospital," said the smiling, mature woman who was sitting behind a glass marked Admissions. It was six a.m., and Jon had to be prepared for his scheduled seven-thirty a.m. surgery. He had been instructed not to eat or drink anything after midnight. Dr. Brad (his nametag said Dr. Bradley Johnson), the anesthesiologist he had met last week, called it "being NPO." Dr. Brad had also told Jon that his laboratory tests were normal.

The admissions nurse handed his chart to another nurse, who asked Jon to follow him. Jon kissed and hugged his mom and dad, as the nurse said to them, "Don't worry, we'll take good care of him." His parents were directed to a waiting area. Jon followed his nurse to the surgical preparation area.

In the preparation area, Jon changed into a hospital gown. He was pleasantly surprised that it fit better than he had heard it would. He was asked to lie down on a very comfortable stretcher. A nurse started an intravenous line in his left arm. This appeared to be her job, as she was going from patient to patient starting them. It was a little pinch, and he soon became unaware of it.

After a short while, Dr Brad came to his stretcher. He reviewed the anesthesia that Jon would have. He told Jon that after Dr. Penski saw him, he would get a relaxing medicine through the intravenous tube. Jon thought that would be great. He had anticipated having to place himself in a yoga trance prior to entering the operating room, but it now seemed that doing so would not be necessary.

In a few minutes, Dr. Penski came to him. She said she had just reviewed the surgery and postoperative course with his parents, and asked if he had any questions. Jon told her that he was looking forward to his new appearance. Dr. Penski assured him that he'd be pleased.

A nurse came by and placed a medicine in the intravenous line, saying it would make him feel good. Soon a feeling of well-being came over him. He had the sensation of the stretcher moving, but he wasn't sure it was really happening.

VII

When Jon's eyes opened, he saw the ceiling of a dimly lit room. He didn't know where he was. He heard the soft hum and beat of modern medical instruments in the background, and he saw people in hospital uniforms moving around quietly and quickly. As his memory came back, Jon realized that he was in the recovery area Dr. Penski had described. His surgery must be over.

"I understand everything went fabulously well," said a female voice.

Jon looked to his left and saw the warm and reassuring smile of a nurse.

"Don't try to talk." She placed an object in his hand. "If you need anything, press this button."

Jon remembered that his teeth were to be held in their new position with tiny elastic bands, and he should not try to open his mouth. At the moment, he couldn't feel his face due to the numbing solution Dr. Penski had given him, so he wasn't sure where his teeth were positioned. Might as well catch up on my sleep, he thought, and immediately fell into a peaceful sleep.

VIII

For the next two days, Jon adjusted to his surgery. He learned to eat—with either a straw or a spoon—the nutritious food that had been processed in a blender. He could talk and be understood, even though his teeth were together, and his speaking was getting better with practice. Although he was taking an anti-swelling medicine, Jon had mild swelling that Dr. Penski said was normal. She told him that she was quite pleased with the results of the surgery.

Jon, on the other hand, was ecstatic with it. Using both a hand mirror and the bathroom mirror, he was able to get a full view of his profile. He was amazed at the difference between his new look and what he remembered of his old one. His lower jaw was still strong, or what would be considered masculine, but it was quite a bit smaller and in a better position than before. He looked at his profile often.

Dr. Penski visited Jon twice a day for the two days he was in the hospital after the surgery. He was discharged the morning of the third day, with instructions to make a postoperative appointment at Dr. Penski's office.

Two weeks after the surgery, Dr. Penski removed the elastics that were holding Jon's teeth together. He was to be on a soft diet, but at least he was able to open his mouth. He was back to his normal lifestyle, except for strenuous sports or similar activities. When he went to his six-week post-surgery checkup, Dr. Penski reviewed his x-rays and pronounced him healed. She didn't need to see him for another three months, for his final visit to her office.

IX

During the weeks following his surgery, Jon was gratified by the reaction of his friends and classmates. Some visited him while he was in the hospital. They had accepted his decision to have the surgery and encouraged him during his healing phase. They were impressed with how great Jon looked. A few of his friends confided that they might make an appointment to see Dr. Penski for problems that they had. Most facets of his life did not change. He was still on the track and swim teams at school, and he remained involved in his beloved karate. What changed was Jon's self-confidence. He was finally pleased with his appearance, and it showed.

One evening at about nine, while Jon was in his room studying history, the phone rang. He picked it up. "Hello."

"Hi, Jon, it's Heather. I was wondering...if you're not busy Saturday..."

Definitions: Jon

Orthognathic Surgery: Reconstructive surgery of the facial bones in order to correct a functional and/or cosmetic defect. This surgery also involves the soft tissues of the face, as the movement of the facial bones results in associated movement of the soft tissues. When asked what the difference is between the oral and maxillofacial surgeon and a plastic surgeon, the classic answer is that the oral and maxillofacial surgeon deals with hard tissue (bones), and the plastic surgeon deals with soft tissue (skin). In reality, both types of surgeons deal with hard and soft tissue, and use what are called plastic surgical techniques.

Osteotomy: Surgery on a bone, such as a facial bone, to reshape it or move it is called an osteotomy.

Facial Bones: (see Definitions: Calista for additional description):

Frontal Bone: Forehead

Nasal Bones: Nose

Zygomatic Bone: Cheek

Orbit: Eye (includes multiple facial bones; e.g., zygoma, frontal, maxilla)

Maxilla: Upper Jaw (including upper teeth)

Mandible: Lower Jaw (including lower teeth and chin)

Sinuses: Some facial bones contain sinuses, or spaces, within them. The spaces help us by lessening the weight of our heads and aid in our speech. (Note the change in your voice when you have a cold or sinusitis.) It is the maxillary sinus with which most people are familiar. When we have sinusitis, it is usually the maxillary sinuses that are infected. Other sinuses associated with the facial bones are the frontal, ethmoid, and sphenoid. When all of the sinuses are infected at the same time, it is called a pan-sinusitis.

Sagittal-Split Technique™: This technique allows the surgeon to perform osteotomies on the mandible from within the mouth (intraoral). Prior to its development, such surgery required incisions on the skin. Even though the skin incisions are quite cosmetic, the intraoral procedure, though technically involved, is an improvement from a cosmetic point of view.

Genioplasty: Refers to surgery of the front (chin) of the mandible. Such surgery is frequently done. It may be part of a multiple bone surgical procedure, or it may be the only surgery performed. The chin can be moved backward, forward, up or down. Bone can be removed or added to it. Implants of various materials are often inserted in the chin.

Predictive Surgery: With the use of computers, x-rays, photos, and plaster models of the face and/or teeth, the surgeon can determine the cosmetic and functional result of a procedure. This is done by doing "pretend" surgery using the aforementioned items. The patient can see his or her present condition and the anticipated change(s) that the surgery will bring.

Auto-Transfusions (Blood): In complicated orthognathic surgical procedures, a patient may require blood replacement during the surgery. In such cases, the doctor will prefer to have the patient receive his or her own blood. While blood transfusions are extremely safe, receiving one's own blood eliminates all chances of rejection, etc.

Auto-Transfusion Technique: The patient is placed on a high-iron regimen and scheduled for auto-transfusions well before the anticipated surgery. On the first auto-transfusion visit, one unit of blood is drawn from the patient and saved. On the next visit, when the patient's body has naturally replaced the previously taken blood, the unit that was drawn is returned to the patient and two units are removed and saved, leaving the patient lacking only one unit. Two units of blood are now available for the patient. If more than two units of blood are deemed necessary, the process continues. The human body tolerates this procedure quite well.

Intermaxillary Fixation (IMF): This is the situation in which the lower teeth are held in occlusion (biting relationship) with the upper teeth. The teeth are usually held in place by tiny elastic bands called intermaxillary fixation elastics. Stainless steel wires may also be used. Patients often refer to IMF as the teeth being wired together. In the past, IMF lasted for a four- to eight-week period. Today, with the help of bone plates (see below), IMF is usually no more than a week or two.

Bone Plates: Made of stainless steel, titanium, or other materials, they are used to hold bones that have been surgically (e.g., orthognathic surgery) or traumatically (e.g., auto accident) fractured. Screws are used to hold the plates to the bone. There are many sizes and shapes of metal plates for use in all parts of the face and head.

Anesthesiologist/Anesthetist: Doctors and nurses who specialize in giving anesthesia to patients. One who has gone to medical school prior to anesthesia training is called an anesthesiologist. One who has a nursing degree and has received specialized anesthesia training is called a nurse anesthetist. Both are equally competent in administering anesthesia to patients, although anesthesiologists are usually in administrative charge of hospital Departments of Anesthesia.

Calista

I

If, from the time you were a little girl, your dream was to be a pitcher on a college fast-pitch softball team, you wanted to be tall when you grew up. It seemed that softball pitchers were taller than most of their teammates. The same would have been true if Calista Morgan had been interested in basketball or tennis. If you were tall, you had an advantage. Both Calista's parents were tall. She knew that when she saw them standing next to their friends. Calista, on the other hand, in her early school years had always been on the short side. No matter how straight she stood, she was shorter than most of her classmates.

Once, someone told her that stretching her body would make her taller. Calista had her dad hang a bar from the ceiling in their basement. Several times a day, she jumped up to it and hung until her arms hurt. After weeks of this, she was no taller than before; her only improvement was in the length of time she could hang. Calista came to the conclusion that the only way she could help herself become taller was by eating right, exercising, and getting the proper amount of sleep. Beyond that, it was up to Mother Nature.

Calista tried out for a pitching spot on her seventh grade softball team. Feeling that she would make a good second baseman or shortstop, the coaches tried to discourage her from pitching. But when they observed the speed and accuracy of Calista's pitches, the coaches realized that she had a special talent. Perhaps it was partly technique and partly a function of her slender body, but Calista threw a softball harder than anyone on the team. She soon became their number one pitcher, and, as can happen in softball, pitched almost every game.

The reason that Calista became interested in softball at an early age was that she lived in Oklahoma City, the home of the women's college softball World Series. If you lived in O.K.C., it was easy to become a softball fanatic. She played pickup games with her neighborhood friends almost every day of summer vacation, and any other time possible during the remainder of the year. Her parents, Jim and Nancy Morgan, often took her to beautiful ASA Hall of Fame Stadium, where the college games were played. The excitement of the games, the noise of the fans, and the smell of hot dogs and peanuts excited Calista. Fans got to see women's softball teams from colleges all over the USA. More than once, the family arrived in the early afternoon and didn't leave until after dark. Dinner was supplied by the concession stands.

Calista never tired of watching the games and hated to see them end. She liked and followed many of the players, but she favored the pitchers. Some pitchers were known nationally, and they had their pictures on the covers of magazines or were seen in television commercials. With the shortened dimensions of softball fields, softball pitchers threw the ball faster—relatively speaking—than Major League baseball pitchers. Games were often pitching duels with few runs scored, and the strategy employed by the coaches was pure joy to those who knew the game.

Throughout her middle school years, Calista did some growing, but remained in the shorter half of her class. Then it happened. It was during Calista's thirteenth year, when she was a freshman at Lynn High School. She began what is commonly called a growth spurt. To her happy amazement, she had grown almost five inches by the time she became a sophomore, and she grew another four inches during that year. Her doctor had never seen such growth in so short a time, but medically Calista was normal. She literally grew out of her clothes. It felt strange to her to have to tilt her head down when she spoke to friends who just a short time ago had been taller than her. By the time Calista entered her third year of high school, she was five feet eleven inches tall. Many college softball pitchers were taller than that, but, height-wise, Calista was in their league. She had arrived!

II

"Gonna throw hard, Calista?" called Sam, the team's catcher, as she ran onto Medford High's softball field, where today's game was scheduled to be played. Calista, who had been in the first of two vans from Lynn High and had arrived a few minutes before Sam, was standing next to third base. These were the first words Sam always spoke to Calista on a day that she was pitching. Sam continued at full speed and stopped just short of her. Someone not familiar with this particular habit of Sam, running and stopping short, would be inclined to jump out of her way. Calista held her ground and Sam stopped a foot or so away. Sam had her catcher's mitt on, and punched it when she talked as a means of emphasizing her point. She was a tad shorter than Calista, and at a solid one hundred and forty five pounds, very powerful.

"I want you to make my hand sting today," she continued, as she punched her mitt. "Since it's the last game of the regular season, I'll have a chance to rest it for a few days."

"For you, anything," retorted Calista. "Anyway, it's a beautiful day for a game." It was a temperate May afternoon, with a mild breeze and puffy clouds for shade. Calista knew how valuable Sam was to the team. She was their best hitter, and a sensational catcher. Many softball insiders felt that the catcher was the most important player on the team. The catcher helped the coaches with game strategy and controlled the pitchers. Nothing made a pitcher happier than having a good catcher.

The other players on her team were approaching the field in groups of two or three. Several were talking animatedly.

"I'm a little nervous," Calista said. "It's an away game, and Medford's beaten me on this field once."

"They were just lucky that day," Sam encouraged. "It won't happen again."

Lynn High's Coach Clyde, having parked the school van that she was driving, approached the field. Two ball girls trailed behind her, carrying bags of bats and softballs. Coach Clyde called for the team to gather around her.

Coach was a plain woman with straggly straw colored hair. She was also an exceptional coach with an envious winning record. She was well-grounded in the fundamental techniques of softball, and knew how to pass on her knowledge to her players. Her claim to fame as a coach, though, was that she was a gentle teacher with infinite patience. Coach also hated to lose.

With her characteristic get-to-the-point manner, Coach Clyde began. "I'm not going to give you a 'rah-rah' speech or anything like that. You're big girls, and you know the importance of this game." She paused a moment to glance around at her players. "We're in a first place tie with Medford; and you know that if we win today, we are regional high school champs and we'll go to the state championships." Another short pause. "All I ask is that you play the way you've been playing. If you do, we'll win. Simple as that." She paused again, prior to her finale. "But remember, playing softball should be fun. That's what this game is all about. So above all, have fun out there today...and kick Medford's butts."

With that preamble, Coach Clyde had her team take the field for pre-game practice. On the other side of the field, Medford High's Coach Gambelli had her team around her and was, most likely, saying words similar to those just spoken by Coach Clyde. They would have their pre-game practice after Lynn High left the field.

About an hour later the game began. Lynn High was at bat, at the top of the first inning. The first batter, a left-hander and a slap hitter, slapped the ball to the shortstop, who made a great play to put the runner out at first. The next batter attempted to bunt for a base hit, but the Medford third baseman anticipated the move and threw her out at first base in a close play.

Sam was the third batter, and hit the second pitch over the outstretched glove of the Medford second baseman. The ball had eyes, and rolled between the right and center fielders to the fence. By the time the right fielder threw the ball in, Sam was on third base brushing dust off her uniform.

Becky, Lynn's left fielder, was next at bat. The Medford pitcher was nervous with Sam at third. Becky swung and missed the first pitch. She did the same with the second pitch. Ahead in the count, but concerned about Sam, the Medford pitcher threw a riseball that sailed high and inside, causing Becky to duck out of the way. The ball accidentally hit her bat and rolled half way to the pitcher's mound. Sam began running as soon as the ball hit the bat and barreled into home plate just ahead of the throw from the pitcher, who had hesitated a bit before fielding the ball. In the confusion occurring around home plate, Becky ran safely to first base. Had the pitcher thrown the ball to first base instead of home plate, Becky would easily have been the third out of the inning. Instead, it was one to nothing, Lynn High.

The Lynn High players en masse came from their dugout to cheer and high-five Sam. It was a few minutes before things settled down enough for Allison, Lynn's first baseman, to come to bat. The first pitch was just what she wanted, and she hit a rope to the left fielder, who didn't have to move a step to catch it.

After that exciting start, the game turned into a classic pitcher's duel, with both pitchers displaying stellar skill. At the bottom of the seventh inning, with the heart of Medford's lineup due at bat, the score still was Lynn High, one, and Medford High, zero. Pitching spectacularly, Calista had given up only one hit, a bunt down the third baseline that refused to roll foul. All she had to do was to hold Medford now, and her team would be the regional champions in softball.

Medford's number two batter, the one with the bunt hit, was up. Sam called for Calista to throw two riseballs followed by a dropball, which had the batter out in three pitches. Sam walked slowly to the pitcher's mound to talk to Calista. "Showed her," Sam said, as she pounded her mitt. She liked to talk tough. Her trips to the mound had a calming effect on Calista. "We'll do the same for Jennie."

Jennie, the next batter, was the best hitter on the Medford team. So far, Calista had held her to groundouts to the shortstop and first baseman. Sam called for a changeup, hoping to catch the batter off balance, as Calista had been pitching fastballs on the first pitch of Jennie's previous at-bats. Expecting a fastball, Jennie was momentarily fooled, but quickly recovered. She hit the ball on the fat part of her bat, creating the metallic sound associated with good contact on an aluminum bat. The ball left the bat on a line toward Calista. There was no time for her to react. With her head turned slightly to the left, as a result of the pitch she had thrown, the ball struck her in the right cheek. The ball bounced up in a loop from Calista's cheek to the shortstop, who instinctively caught it for the second out. Calista looked like a puppet whose strings had suddenly been cut. She crumpled to the ground at the front of the pitcher's mound. Players and coaches from both teams ran to her aid. They gathered around her, several shouting, "Give her air!" Confused, Calista looked up and asked, "What happened?"

Calista's nose was bleeding and a bruise was beginning to form on her right cheek. She attempted to get up, but she was told not to move. Coach Clyde called 911 from her cell phone. Next, she called Calista's mom to inform her of what had happened. (Coach got the number from the emergency contact list she had for each player.) She told Mrs. Morgan what had happened and that Calista was conscious and responsive, but that she would be transported to the hospital for an evaluation. The ball girls quickly filled plastic bags with ice and placed them against Calista's cheek. A wet towel was pressed under her nose to stop the bleeding. A towel was gently placed on each side of her head to limit movement. Ten minutes later, the ambulance arrived.

The paramedics quickly evaluated Calista's condition. After confirming that she was alert and oriented, they started an intravenous line, stabilized her spine, and placed her in the ambulance for transport to Day Memorial Hospital. Throughout all this, Calista lay quietly without complaint.

III

Calista was admitted directly to the emergency room. A nurse and doctor determined that, most likely, her only injury was to her cheekbone; nevertheless, x-rays were ordered to evaluate her head, spine, and facial bones.

After a while, Calista was placed in one of the side rooms of the emergency department. The door was a curtain that made it easy to call for help, if needed. Her mom was sitting by her side and holding her hand. Dr. Hubert, the emergency room doctor who had initially evaluated Calista, entered the room.

"Hello again," he began, looking at Calista. "I've reviewed your tests and x-rays, and everything is fine except for your cheekbone. You've probably cracked it." Dr. Hubert waited for a response.

Mrs. Morgan nervously said, "My God. What does that mean?"

"It means, Mrs. Morgan, that I think one of Calista's facial bones—the one that we call the zygoma—has separated from its attachments and needs to be repositioned. This is not an uncommon injury; we often see this type of injury in auto accidents. But before I go any further, I want you to know that I've called an oral and maxillofacial surgeon to evaluate and treat Calista. His name is Dr. Barney Ross. While I can't speak for him, I'm sure that he will want Calista admitted to the hospital, to ensure that she is stable. Perhaps tomorrow, if he confirms my diagnosis, he will take her to the operating room and fix her up. She should be able to go home later tomorrow, or the next day."

Calista, a so-called tough cookie, seemed to take what had been said quite well. Her mother, on the other hand, was not as tough. Her reaction to what Dr. Hubert had just said was to slump back into her chair and appear close to fainting. A nurse quickly brought a cold cloth for her forehead. After ascertaining that Mrs. Morgan was fine, Dr. Hubert left the room.

IV

The cloth was still on Nancy Morgan's forehead, when the curtain opened and a face peeked in. "Hello. Are you Calista?"

Both Calista and her mom said yes.

Stepping into the room, the new arrival said, "I'm Dr. Ross, Barney Ross, an oral and maxillofacial surgeon. Dr. Hubert called for me to come and see Calista." Dr. Ross was tall and slender, with a long, narrow face. He might have played Sherlock Holmes© in a 1940s movie. He was holding a large cardboard packet of x-rays. "Oh, I've been instructed by a group of excited girls in the waiting room to tell you that your team won the game, and you are regional champions."

"Great!" Calista said, appearing genuinely happy at the outcome of the game in spite of her injury.

"I also told them that you were going to be fine. That made them very happy."

Having somewhat regained her composure, Mrs. Morgan stood and extended her hand to Dr. Ross. "I'm Nancy Morgan, Calista's mother. I'm so happy that you're here to take care of my daughter."

"I'll do my best, Mrs. Morgan. Have a seat, and I'll examine Calista." He waited for Mrs. Morgan to be seated. "As Dr. Hubert told you, and as my review of Calista's x-rays indicates," he said, looking directly at Calista, "you, most likely, have a cheekbone fracture."

Calista nodded her head in agreement with what Dr. Ross had said. Dr. Ross put the x-ray packet down in preparation of his examination of Calista, who continued to lie calmly on the gurney. Dr. Ross was aware that Calista had been given a complete examination upon admission to the emergency room; therefore, he limited his exam to her head and neck region. He looked at her skin and noted the bruising and swelling of the right cheekbone. He checked her hair and scalp for injuries or lacerations, and saw none. Dr. Ross placed his index finger at the outside of Calista's right eyebrow and pushed a little. "Does this hurt?"

Calista winced. "Ouch!"

"Sorry." He then placed the same finger and ran it under her right eye. He found an area near her nose and pushed again. "Does this hurt?"

"A little."

Dr. Ross' expression was noncommittal. Taking a small flashlight from his left breast pocket, he ran its light across the pupils of Calista's eyes. Dr. Ross noted that both pupils were equal, and reacted to the light by momentarily constricting. This light test was an evaluation of Calista's neurological status. She passed with flying colors.

Shining the light at the whites of her eyes, Dr. Ross saw redness caused by blood at the outer edge of the right eye. He had Calista follow his finger with her eyes: left, right, up, and down. Dr. Ross told Calista that this test was to find out if the muscles of her eye were what he called "entrapped." If her eye muscles were entrapped, they would not move in their normal coordinated fashion, resulting in her seeing double when she looked in certain directions. The redness was the only abnormality he noted with her eyes, but Dr. Ross knew it was a sign that the right cheekbone was broken.

Dr. Ross shined his light up the nostrils of her nose and saw no bleeding, only crusted blood. Then, he took a cotton swab and lightly touched her skin with it. Calista said the area under her right eye was tingly and numb. As with the redness in the outer corner of the eye, the numbness was a sign of a fractured cheekbone. The examination was completed with an evaluation of Calista's neck. There seemed to be no problems in that area. At the conclusion of his evaluation, Dr. Ross took a step back, leaned against the wall, and folded his arms. "My examination confirms what the x-rays indicated."

"The cheekbone?" asked Mrs. Morgan.

"Yes, Calista has fractured her right cheekbone."

"Fractured? Is that different from broken?" continued Mrs. Morgan.

"No," replied Dr. Ross. "Fractured and broken mean the same thing, just as hemorrhage and bleeding mean the same thing. People generally think that these terms imply different conditions, but they don't." Dr. Ross took a moment for Mrs. Morgan and Calista to react to what he had said. They didn't seem to have questions. "I'd like to explain Calista's injury to you. If I become too technical let me know."

Both Calista, who was paying attention from her gurney, and Nancy Morgan nodded.

Dr. Ross continued. "When Calista was hit with the softball, it caused her right cheekbone to become detached from its surrounding bones. The cheekbone is officially called the zygoma or malar bone. There are two of them, one on each side. They are two of the many bones that come together to form the face. I could feel the areas of the fracture when I pressed under your eye and at the outside of your eyebrow. They're called 'step defects,' and are two of the areas where the zygoma bone is attached to other bones. If I don't fix it," he stated, looking at Calista, "you will have a deformity and possibly a vision problem."

Both Calista and her mom began to speak. Calista was a bit quicker. "What kind of problems?"

"Your face will look flat on the right side, and your right eye may be displaced lower and be somewhat sunken in. You may also have double vision when you look in certain directions. Let me show you the x-rays."

Dr. Ross opened the packet of x-rays, which were really digital photos, the modern replacement of the classic x-ray. He showed Calista and her mom the areas of fracture, pointing out the step defects he had mentioned. When he was sure that they understood the details of the fractured cheekbone, he put the x-rays away.

"Wow!" was the comment from both mother and daughter, and then it was Mrs. Morgan's turn. "Can you fix it?"

Dr. Ross smiled. "I certainly can. Let me tell you what we need to do."

Calista and her mom listened attentively as he continued. "Since you've eaten recently, and your fracture doesn't require immediate treatment, I'd like to formally admit you to the hospital and have you spend the night. You can relax, get pain medicine if you need it, and even eat a bit—until midnight, that is. After midnight you'll need to be NPO. That's a fancy way of saying you can't eat. I'm also going to continue your intravenous fluids. That way, I can ensure that you'll get the fluids and some nutrients you need. Any questions so far?"

Mrs. Morgan said, "My husband, Jim, is out of town on business. Do I need to have him come home?"

"No, Mrs. Morgan, we'll do fine without him. Tell him not to worry," replied Dr. Ross. Turning toward Calista, "In the morning I'll take you to the operating room, where you will be given general anesthesia—that is, you'll be put to sleep—and I'll fix your fracture."

Calista looked nervously at her mom, who knew what Calista wanted to know. Mrs. Morgan asked, "Just how do you fix her cheekbone?"

Dr. Ross thought for a moment, and then said, "Watch my finger." He ran his right index finger along the upper part of his right eyebrow, from his nose to the outer corner of his eye. He stopped his finger at the outer corner of the brow and held it there. "My finger is at what we call the lateral brow area. It's where I pushed before, and it hurt."

"I remember," commented Calista.

Dr. Ross said, "Right here, just above the eyebrow, I have to make a small incision, about an inch or so long. It's a very cosmetic incision, and when it heals completely, you'll have to look hard to find it." He paused to judge their reaction to what he had said. Both listeners reacted well. "Once again, if I'm too graphic, stop me. I'm then able to pass an instrument through the incision line, allowing me to manipulate the fractured zygoma into its original position and stabilize it with a stainless steel wire near the incision area. If one wire is not enough—and it usually is—I will need to make a second incision immediately under your lower eyelid, and place a wire in that area." He ran his hand under his lower lid to demonstrate the area where a second incision would be made.

Calista raised her hand to indicate she had a question.

"Yes, Calista?"

"First, I like the way you tell it like it is, Doc."

"Thank you," replied Dr. Ross.

"But won't I feel the wires? Do you have to take them out later?"

"Except in rare instances, the answer to both questions is no. The wires are very thin and are hidden under the tissues. You will be unaware of them. There are many people who walk around with wires or plates in their bodies. You won't even cause the metal detector at the airport to go off."

This latter remark evoked chuckles from both mother and daughter.

"In the very unlikely circumstance that you need a wire removed, it's a relatively minor office procedure to remove it."

"That's reassuring," said Mrs. Morgan.

Calista anxiously raised her hand again, signaling another question.

"Go ahead, Calista."

"How long will it take for my cheekbone to heal, and when will I be able to play softball? I know I'll miss the state high school championships, but the summer league will be starting soon, and I want to play with the O.K.C. Cyclones."

Dr. Ross considered this, and then said, "At your age, your bone will be healed in about four weeks. I think you should not play softball for about two weeks or so—or at least until any stitches I place are removed. After that, you can practice light duty, if such a thing exists in softball. Within a month or so you can return to your normal softball routine."

"Stitches!" Calista hadn't thought about the need to repair the incision line.

"Don't worry, Calista. They're very fine...thinner than your hair, in fact. I take them out within a week. It takes just a few seconds to remove them. In the meantime, they are hidden under a small bandage."

"Listen to the doctor, Cali," her mom encouraged. "He doesn't want to hurt you."

"You're right, Mrs. Morgan. I want Calista to have the best experience that I can provide." Dr. Ross paused, then looked at Calista and said, "You will miss just a few days of school. I hope that doesn't disappoint you." He smiled.

"As I said before, thanks for telling it like it is, Doc." Calista put her head back down on the stretcher, indicating that she was ready to get on with her treatment.

"Calista! Be more respectful to the doctor."

"No problem, Mrs. Morgan. I'm favorably impressed with your daughter's attitude. Do you have any questions?"

"No, Doctor. You've been very clear about Cali's injury, and how it needs to be treated."

"Okay then," said Dr. Ross as he straightened up, "I'll let the emergency room staff know that you're ready to be sent to your hospital room. Have a good night's sleep. I'll see you in the morning." He began to leave.

"Bye, Doctor," said Mrs. Morgan.

"See you," said Calista.

V

The next morning Calista was given a sedative while still in her hospital room. By the time she arrived in the operating suites area, she was quite relaxed and unconcerned about the upcoming surgery. She remembered being greeted by the anesthesiologist and informed that she would be putting Calista to sleep, but nothing else until Calista awoke in her room after the surgery.

Her mom was sitting in a cushioned chair by the side of her bed. It took a while for Calista to assess her situation. Surprisingly, she was in no pain. Touching the area of surgery with her hand, she noted there was a large bandage wrapped around her head that was quite thick over her right eyebrow. There seemed to be a piece of wood sticking out from it. Calista's first thought was that this didn't seem to fit Dr. Ross' description of a small bandage over some stitches. At the moment, she was too tired to be concerned. Calista snuggled into her pillow and fell asleep.

She awoke a few hours later feeling surprisingly refreshed... and hungry. Her mom was still sitting—or rather, napping—in the chair by her side. Calista found the nurse call button and pressed it.

A nurse appeared almost immediately. "Can I help you, Calista?"

"Yes, I'm a little hungry. Is there anything to eat?"

"Great, I'll have something brought to you. By the way, do you feel like going to the bathroom? We need to have you go, before you go home."

"Now that you mention it, I think I can." The nurse escorted Calista to the bathroom. After Calista had gone to the bathroom and eaten a bit, her intravenous line was removed. At about 2:00 p.m., Dr. Ross arrived in her room. He nodded hello to Mrs. Morgan.

To Calista, he said, "How are you feeling?"

"Real good. But what's with this pile of bandages and piece of wood I have around my head? I thought you said I would have a small bandage."

Dr. Ross chuckled. "You're right. And you will have a small bandage. What you have now is my surgical dressing, which I'll remove tomorrow. It will protect the area of surgery for twenty-four hours or so."

"I didn't mean to question you, Doctor. I was just wondering."

"Please, Calista, feel free to ask any questions you like. That's what a good patient does. Since you've gone to the bathroom and eaten, I'm going to discharge you from the hospital. I want you to take it easy today. I'll see you in my office tomorrow morning, when I'll take that bulky bandage off and replace it with the small one I promised you."

"That's great!" Calista literally shouted. "This wasn't as bad as I thought it would be."

"Yes," her mom said, "we all thank you for your fine treatment of Cali. Jim said to tell you that he is so pleased with all you've done."

Dr. Ross replied, "I appreciate your kind words. Thank you, and I'll see you tomorrow," he said, and left the room.

VI

A week later, Dr. Ross escorted Calista and Mrs. Morgan from the treatment room in which he had removed the last of Calista's stitches. Calista had noticed that the office was a bit old-fashioned, with vintage furniture. The wall coverings were combinations of light and dark brown, and prints of old masters were on the walls. All in all, it seemed to fit Dr. Ross.

"Everything looks fine. You appear to be healing well. Leave the bandage strips that I placed across the incision line in place for three days, and then gently remove them. You can do that by taking a shower and getting them wet. They should be easily removable at that time."

"I'll do that. See you in a month."

"Sooner, if you have problems. And remember, no tough softball practices for the next four weeks."

"Okay, Barney—I mean, Doc—I mean, Doctor. I'll be good, if you insist." Calista couldn't help laughing at her attempt at humor. She felt that she knew Dr. Ross well enough that her little joke wouldn't insult him.

VII

It was a little less than two months since her injury. School was out; the summer softball league had started, and today was to be Calista's first time pitching in a game since she had been injured. She had followed Dr. Ross' instructions and had begun her return to softball with light practices, which within a few weeks graduated to full practices. Calista was physically at or above her pre-injury state, but her mental status remained questionable. Although she had looked forward to this day, she had to admit that she was a bit nervous. Would she be afraid to step on the pitching rubber and face a batter who might hit a ball right back at her? Had she lost her courage? She didn't think so, but wouldn't know for sure until after the game began.

As usual, Sam came barreling toward her, stopping just short of running her down. "Gonna throw hard?" Sam's habit brought comfort with its familiarity and chased away some of her jitters. Because it was a hot and sunny Oklahoma summer day, Sam was breathing fast and had sweat above her upper lip. Her uniform shirt, already a bit wet from perspiration, was untucked—as usual—giving the impression that she was even bigger than she was. The other members, including Calista, of the Oklahoma City Cyclones wore their shirts neatly tucked in. Sam said that getting into and out of her catching position automatically untucked her shirt, so why not always wear it that way?

"I'm a little nervous, Sam."

"Look, you wouldn't be normal if you weren't nervous." Sam took a step back, put her hands on her hips, then hit her mitt with her fist and looked directly into Calista's eyes. If Calista didn't know better, she'd have thought that Sam was going to spit the juice from a plug of chewing tobacco before she spoke. "Let me tell you about my cousin Bobbie." Calista had heard Sam talk of her cousin Bobbie, but had never met her. Sam continued, "Bobbie rides horses. She hopes to compete in the Olympics, someday, in the equestrian events. Two years ago she fell off her horse while practicing jumps and broke her collarbone. That was a tough time for her. For a while, she thought that she might not ride again." Sam paused for emphasis and then continued. "When her collarbone healed, Bobbie was scared, yes—and it took all her courage, yes—but she got on her horse and jumped a fence." Sam turned her head left and right before focusing on Calista, as if to emphasize her next words. "That's what it took, Calista! Today, she rides and jumps better than before her fall."

Sam paused, trying to determine if Calista understood her point. Calista was at a loss for words.

"Don't you see? That's what you need to do. Get on that mound and pitch! The next time a ball comes right back at you—and you know one will—you're going to catch it." Having said her piece, Sam threw her arms around Calista and gave her a hug of encouragement. "See you in the dugout." Sam turned and ran toward the dugout.

Calista stood quietly. There would probably be no other

time in her life when she would so clearly understand the meaning of friendship. Maybe she couldn't put the definition of friendship into words, but Sam had just demonstrated to her what it was. Suddenly, Calista knew she had the courage to face the opposing team's batters. So what if a ball was hit right back at her? As Sam said, she'd catch it. With a newfound confidence and a bounce in her stride, Calista followed Sam to the dugout.

Definitions: Calista

Facial Bones: Facial bones provide the basic structure of the face. Soft tissues (muscles, skin, etc.) cover and are attached to the bones, either directly of indirectly. The combination of hard and soft tissues give rise to the face that we see when we look in a mirror. The major facial bones are the mandible (1), maxilla (2), zygoma (2), nasal (2), frontal (2) and spheniod (1). Structures such as the sinuses, mouth, eye cavities (orbits), and nasal cavity are within or the result of the coming together (articulation) of the facial bones.

Cheekbone Fracture: The zygoma bone, also called the malar bone, is attached to other bones in four places. Correction of the fracture depends upon the exact category of cheekbone fracture (see various approaches below.) Another name for a cheekbone fracture is tripod fracture.

Loss of Feeling: Anesthesia and paresthesia are two of the terms used to describe numbness. When a facial bone is fractured, an area of numbness of the face may occur that is often diagnostic of the fracture. In the case of a cheekbone fracture, numbness of the area under the eye is present. The numbness is the result of injury to or compression of the infraorbital nerve that gives feeling to the area under the eye.

Fracture Reduction: The fixing of a fracture is called a reduction. In simplest terms, there are two types of reduction: an open reduction and a closed reduction. In an open reduction, an incision is made to expose the fracture site, and wires and/or plates may be utilized to stabilize the fractured bone while it heals. In a closed reduction, no incision is made and the bone is immobilized during the healing period by such means as wiring the teeth together (intermaxilly fixation).

Lateral Brow, Infraorbital/Intraorbital, and Gillies Approaches: Skin incisions commonly utilized in the reduction of a fractured cheekbone. In a lateral brow approach, an incision is made just above the outside edge of the eyebrow. In an infraorbital/intraorbital approach, the incision is made just below or just above the lower eyelid. In a Gilies approach, an incision is made in the hair in front of and above the ear. The choice of approach depends upon the category of cheekbone fracture.

Trapdoor Fracture: It is not uncommon that the paper-thin bone under the eye (the floor of the orbit) is also fractured when the cheekbone is fractured. In these situations, the bone protrudes down into the sinus below it, reminiscent of an open trapdoor. The eye muscles and the eye itself may drop into the trapdoor opening, causing vision and cosmetic problems. The infraorbital/intraorbital approach provides access to this fracture site, allowing for the placement of an implant to replace the orbital floor.

Redness of the Outer Aspect of the Eye: Because of the anatomy of the eye and its attachment to the orbit (the bony house of the eye), erythema (redness) of the outer aspect of the eye is commonly found in cheekbone fractures. The redness helps in diagnosing such fractures.

Toby

I

You had to say that Toby Stevenson was what you would call gifted. For as long as anyone could remember, there was no intellectual endeavor he attempted in which he was not successful. For example, when he was four years old, he decided to learn to play the piano that was sitting unused in his family's living room. Seeing Toby's enthusiasm for the piano, his parents hired a teacher to come to their home and teach him. Toby was such a quick learner that after a few years, the teacher excused himself, saying that Toby had exhausted his repertoire. He suggested that a more advanced tutor be employed, perhaps one associated with a professional institution such as Tanglewood. Toby declined the suggestion, as he preferred self-teaching. His parents acquiesced to Toby's preference, and he eventually became an excellent pianist. At nine years old, he was playing on weekends in an exclusive department store in Boston, his hometown. Toby remembered how proud he had been when shoppers gathered around his piano and admired his playing. He played much less now, but he still played wonderfully.

In high school, he excelled in his studies, especially mathematics. Toby had read somewhere that there was a relationship between music and mathematics, as they were two fields where true prodigies were found. He didn't think he was a math or music prodigy, but he knew some might consider him one, especially in math. When he took the college SATs, Toby aced the math part—to the extent that he was a candidate for institutions such as MIT, Stanford, and Cal Tech. He knew that he would be extremely fortunate to be a part of any of these places of higher learning.

II

After much mental debate, Toby chose and was accepted by MIT. The fact that he wouldn't have to leave his beloved Boston was a large factor in his decision. Toby's initial encounter with the institution led him to believe that the freshman curriculum was based on the assumption that students were prepared to work hard. Days that started at 5:00 a.m. and ended at midnight were grueling. His consolations were that he got excellent grades in his courses, and he was gaining an amazing amount of knowledge.

One spring day near the end of his first year, Toby happened to be sitting on a narrow stool in the physics lab, attempting to refine his knowledge of laser light. His roommate, Brad Foster, was his lab partner. Brad was as different from Toby as a roommate could be. He had grown up on a farm in Iowa and was a country boy through and through. Toby chuckled when Brad used countrified expressions such as "lying back in the weeds" when he wanted to say that he was relaxing. Brad was also more athletically inclined than Toby. At the present time he was into bodybuilding, with the intention of entering contests during the coming summer break. Toby was impressed with Brad's work ethic. After a hard day of classes and lab work, Brad would return to his room after dinner, study for four or five hours, and lift weights until all hours of the night. On weekends, Brad worked out twice a day. The hard work was paying off, as Brad was doing well in his studies, and his five feet nine frame was packed with muscle. Add to this a handsome face, blue eyes, and blond wavy hair, and it seemed that Brad had it all.

"Hold on there, buddy!" Brad hollered catching Toby by the shoulders as Toby was sliding off the stool.

"Uh...thanks, Brad. I guess I dozed off a moment. If you hadn't caught me, I might have cracked my skull on the cement floor. I guess I need another cup of coffee, or I'll fall asleep again." Toby started to walk to the coffee machine in the small closet they called the break room.

"How many cups have you had already?" Brad stepped in front of Toby, preventing him from heading for the coffee. Even though Brad was shorter than him, Toby knew it would be senseless to try to get by Brad. Several of the students in the room turned to see what was happening at Toby and Brad's lab station.

"I don't know. Five or six...I guess."

Brad looked serious. "Don't you know that drinking too much coffee is bad for your heart?"

"Yeah, but I need something to keep me awake."

Brad's face took on a conspiratorial look. "I've got just the thing for you."

"What's that, Brad?"

"Snuff."

"Snuff? What you use?"

"Yeah, the very same. How do you think I do it? The snuff or smokeless tobacco, whatever you want to call it, gives me enough of a jolt to allow me to keep my crazy schedule. I wouldn't be able to do it otherwise."

"Look, Brad, I like you and all that, but your dipping snuff is disgusting. Your lower lip is always puffed out like someone hit it."

"Listen a minute," Brad insisted. "I felt the same way when my high school buddies put me onto snuff dipping, but once I tried it, I liked it. It's like a shot of adrenaline. The only bad side is that it burns the inside of your mouth a little. In your case, it'll keep you awake so you can do what you need to do. I'm not saying for you to give up coffee altogether, but dipping a little snuff will limit your need for it."

Toby looked and felt uncertain about what Brad was saying. He had heard that smokeless tobacco was not good for the body—much worse than coffee—but Brad was contradicting this. He was inclined to reject the idea of using it, although he didn't want to disappoint Brad, who seemed to fervently want him to try the product.

Stalling for time, he asked, "How do you use it?"

"It's easy," Brad said, taking from his pocket a small plastic can that looked for all the world like a typical shoe polish tin. He twisted the top to open it, exposing a mass of dark, grainy tobacco within. With his fingers, he removed a small amount of it, a pinch he called it, from the container. "See," he said, and placed it in his mouth between his lower lip and gum on the right. "You just hold it in your mouth for a while, like this." Brad pointed to his lip, as if Toby didn't know where the snuff was in Brad's mouth. "Come on…try it."

Toby felt as if he were being backed into a corner. He liked and respected Brad. In fact, Brad had become a role model for him. Because he didn't want to disappoint Brad, Toby took Brad's suggestion and tried some.

It didn't taste good, and he had small tobacco particles all over his mouth, but within a few minutes of trying the smokeless tobacco, Toby felt himself becoming more alert. Brad said that the particles were the result of his not yet perfecting the technique of using the tobacco. After this introduction of Toby to the world of snuff dipping, they went back to their lab station and completed the day's project.

III

Since that day in the lab when he first used smokeless tobacco, Toby had become a frequent user of it. He rationalized that it gave him the same burst of energy that coffee did and was much more convenient. The tobacco looked like coffee grounds; a fact that, at least subconsciously, made him consider the use of it as simply a coffee replacement. In reality, though, Toby was dipping snuff much more frequently than he drank coffee. He was using it not only to energize, but also to relax when he was wound up. Toby found that when he was without the tobacco for more than a few hours, he developed an agitation that was only relieved by putting some tobacco into his mouth. He was intelligent enough to realize that he had developed a bit of an addiction, but nothing—he was certain—he couldn't break if he wanted to.

The concern that was becoming increasingly more important in Toby's mind was the warning on the can of smokeless tobacco that said its use might cause mouth cancer. While he felt that his chances of developing mouth cancer were slim considering his young age, reading the warning sometimes sent a shiver down his spine. Brad was still using snuff as much or more than he, yet Brad seemed not to worry about getting mouth cancer. "If I get it, I'll just have it removed," Brad would say with a chuckle. But still, in the far reaches of his mind, Toby was scared.

IV

In the late winter of their second year at MIT, Brad, who was still Toby's roommate, was shaving one morning. He called Toby into the bathroom. When Toby entered, he said, "Take a look in my mouth and tell me what you see." Brad pulled his lower lip out, exposing the general area where he typically placed the snuff. Toby was taken aback by Brad's request. He wasn't used to looking in other people's mouths.

"What do you want me to look at?"

Brad pointed inside his mouth with his index finger. "There's a sore inside my mouth. I want to know what you think."

Toby wasn't sure he would be able to help Brad, but he decided to try. He leaned over so that he was looking down into the area in question. "Move to the left a little, so I can see better."

What Toby saw, and told Brad, was that there was a vertical slit with thick white skin around it. "You should probably have it checked by a dentist. It's the exact area where you put your snuff."

Brad was visibly shaken. They both thought of the message on the can that warned of oral cancer.

"You're right; I'll see a dentist as soon as possible. In fact, I'm going to call one right now." Brad left the bathroom to make his call.

Toby turned to the mirror and pulled his lip out the see the area where he put his snuff. He was shocked to see that the area was red, with white patches scattered throughout. He realized that he also needed to see a dentist.

V

That afternoon, both Toby and Brad were seen in the screening clinic of a nearby dental school. The student dentist had called an instructor over after she looked in their mouths. The instructor seemed quite concerned, and recommended that biopsies of the irritated tissue in their mouths be taken. The instructor mentioned that the condition might be dangerous. They were scheduled to have the biopsies done the next morning by the attending oral and maxillofacial surgeon.

As they left the downtown building that some would call a skyscraper, Toby was clearly distraught. "I wish I never tried snuff dipping. I'm really worried."

Brad, who, if anything, was more distressed than Toby, said, "You're worried! You've been dipping for only a year or so. I've been doing it for more than four years. I'm downright scared. What if I have cancer? What'll I do?"

In an attempt at machismo, Toby replied, "If we have it, we'll deal with it." He didn't feel much better, and he remembered Brad's bravado when Brad had flippantly said that if he had cancer he would just have it removed.

The next day at 9:00 a.m., they arrived at the registration desk of the dental school. A nurse asked if they wanted to be interviewed by the oral and maxillofacial surgeon singly or together. They chose together. The nurse led them to an area of dental suites where, apparently, the private office of one of the doctors was. They sat anxiously on straight-backed cushioned chairs, quite literally twiddling their thumbs. The nameplate on the desk read Marisol Cruz, D.D.S., M.D. Underneath the name, in somewhat smaller letters, was Oral and Maxillofacial Surgery.

After a few minutes, Dr. Marisol Cruz entered the office. Smiling, she said, "Hi, boys. I'm Dr. Cruz, an oral and maxillofacial surgeon. The attending dentist at the screening clinic asked that I see you. They said your names are Brad Foster and Toby Stevenson." She spoke with a slight Latin accent, and didn't appear much older than they were. She was about five feet seven and rather slim. Her most prominent facial feature was her mouth. It was evident that she liked to smile, and she had a smile that could light up a room. Brad stood and held out his hand. "I'm Brad, and that's Toby," he said, pointing to Toby. They shook hands.

Dr. Cruz made her way behind his desk, sat down, and faced the boys. She looked first at one and then the other. She began with small talk. "I'll never get used to this cold weather. On days like this, I miss Miami." Both Toby and Brad nodded their heads in agreement, although neither had ever been to that city. They were hoping Dr. Cruz would get to their problem without further delay. As if reading their minds, Dr. Cruz's face turned serious.

"Even though privacy usually dictates that I speak to you individually, you've agreed to allow me to speak to you in each other's presence. That makes sense, as I understand you're roommates and close friends. Also, since you have similar problems, my job is easier. I can say what I have to say only once."

Toby and Brad leaned forward to catch every word Dr. Cruz said. "I'll start with my 'No Snuff Dipping' lecture. Bear with me. I feel it's important that you hear what I have to say. All tobacco, in whatever form, is bad. From this moment, you must stop your use of smokeless tobacco. Cigarettes and cigars generally cause problems with your lungs, while pipes and smokeless tobacco often cause mouth problems. As I'm sure you know, there is a long list of potential problems that these materials cause, from minor irritation to cancer."

At the word cancer, both boys fidgeted and adjusted their seating positions, which Dr. Cruz noted. Pressing on, she said, "Your particular problem is that snuff dipping has resulted in areas of irritation in your mouths. Technically, from your initial screening reports, I would describe Brad's situation as an ulcer with leukoplakia and erythema. That's doctor-speak for a sore with white and red areas around it. Toby, you have the same problem, but without the sore. These areas can be dangerous and must be biopsied in order to determine their exact nature. Once we know what they are, we can treat them, if necessary. Any questions at this point?"

Both patients shook their heads.

"Let me tell you what a biopsy is," Dr. Cruz said, without hesitating. "I'll give you a little local anesthesia, and then I'll take a small piece of the areas in question and send them to the laboratory for evaluation. A stitch that dissolves will be placed in the biopsy site. It will take a few days before we get the results."

Brad had a question. "What if the news is bad? I mean, what if it's cancer?"

"First of all," Dr. Cruz said, "let's not jump the gun. Most often, these things are not cancer. But as I said, if either of you has cancer of the mouth, we'll treat it. I would go into much more detail regarding treatment at that time."

Dr. Cruz folded her arms in a manner indicating that she was finished talking. Seeing that neither Toby nor Brad seemed to have other questions, she stood up and said, "Let's get on with it."

The boys were escorted to the surgical area of the clinic where Dr. Cruz did the biopsies she had described. "Incisional biopsies" she called them, explaining that only a part of the lesion was removed for examination. A small piece of gauze was placed in each patient's mouth, and an appointment was made for a follow-up visit in a week, to review the results of the biopsies. Finally, they were discharged.

Neither spoke until reaching the parking garage. Toby said, "That wasn't so bad."

Brad turned to Toby and asked, "Are you going to tell your parents?"

"No...only if the news is bad."

"Me, too," Brad replied. "God, I'm not looking forward to the next visit."

As Toby continued walking, he thought: Nor am I...Nor am I.

VI

A week later, Toby was once again sitting in the reception area of the main dental clinic, only this time he was sitting alone. Brad and he had arrived together, but they were being seen separately. When the nurse had come for them, they both got up to follow her, but she had informed them that today Dr. Cruz would see them one at a time. Both felt that the separation probably meant bad news. Brad went first, and already had been gone for over half an hour. Toby was becoming more anxious with every passing minute. At last, the door to the private offices opened, and Brad came out.

Toby saw that Brad's eyes were red, as if he had been crying. He immediately got up to ask what had happened, but Brad brushed by him, saying, "Not now...I've gotta go."

Toby was about to say that Brad should wait for him, so Toby could drive them both back to the dorm, but Brad anticipated him and said, "I'll get a cab."

It didn't take a genius to guess that Brad had gotten bad news—in fact, the worst news. Toby felt immense sorrow for Brad, and wondered if he would share Brad's fate, and be given the same diagnosis. Toby suddenly began to shake uncontrollably; a little at first, and then more. He couldn't stop, no matter how hard he tried. In fact, the harder he tried to stop shaking, the more he shook. The nurse coming to take him to Dr. Cruz's office saw what was happening and ran to him. Other medical personnel noticed, and in less than a minute there were four people around Toby. He was gently placed on the floor, a slender tube providing oxygen was placed in his nose, and he was connected to portable monitors that assessed his physical status. It was soon apparent that Toby was fine physically, but he was obviously having a panic attack. Toby was lifted onto a wheelchair and brought to what the staff referred to as a recovery room. A nurse remained to watch him. The room was dark and quiet, and Toby closed his eyes and soon began to feel better. As he was sitting there, Dr. Cruz entered the room and quietly approached. She placed a reassuring hand on Toby's right shoulder.

"How are you doing, Toby?" she asked, in little more than a whisper.

Toby opened his eyes and turned toward the voice. "Hi, Dr. Cruz. Sorry to cause so much trouble."

"No problem, Toby. This happens every once in a while."

There was a pause, as Toby appeared to want to say something. Finally he asked, "What happened to Brad? Why did he run out of here?"

"That's private information that I can't give you. I think Brad will tell you, though. What's more important at the moment is my reviewing your biopsy report with you. When you are up to it, I'll have my nurse escort you to my office. See you in a little while." Dr. Cruz turned and left the cubicle.

Fifteen minutes later, Toby was sitting in front of Dr. Cruz. This time there was no talk of the weather. "Have you been able to stop your snuff dipping habit?"

"I'm trying, Doctor, but as I'm sure you know, it's difficult. I've used it a few times since I was here last."

"I know how hard it is to stop, but you have to stop before it's too late. You are a lucky person, Toby. Your report revealed the beginning of cancer development in your mouth, but it's not what we call 'frank cancer' at this time. You have mild dysplasia, which means you are knocking at the door. Some might say it's the beginning of cancer."

Toby appeared to want to ask a question, but Dr. Cruz held up her palms to indicate that she didn't want to answer questions at this time. "You must have a complete excision of the lesion in your mouth. The procedure requires grafting that utilizes tissue from your mouth or skin. Where we get the graft depends on how much tissue we need to cover the area of surgery. If I find that I don't need a great deal of tissue, I'll take it from your mouth. If I need a lot, I will have to take it from your hip, inner arm, or a similar area. Admission to the hospital for one day is required. It will be a few weeks before you completely heal, and during that period you will be on soft foods and medicines that I will prescribe. Of course, you'll miss some of your classes. How many? I can't say. It depends on how you feel. Okay, now you can ask questions."

All Toby could say was, "Wow!" He didn't say another thing for about a minute, although it seemed longer. Dr. Cruz said nothing, and waited patiently.

"What if I don't do what you recommend?"

Dr. Cruz thought for a moment before saying. "In my opinion, that would be unwise. There is a strong possibility that the dysplasia will progress and become cancer, especially if you continue snuff dipping. Treatment of cancer is much more involved, and—as you know—cancer can kill." Dr. Cruz became forceful, "Look, Toby! You really have no choice in the matter. You have to do what I recommend, if you want to put this matter behind you."

Toby stared down at his shoes. All he could say was a soft, "Okay." The surgery was scheduled for the following week.

VII

When Toby returned to his dorm room, the lights were off. He found Brad sitting quietly in a corner and staring out the window. Brad didn't acknowledge him. Toby wanted to say something—anything—to comfort Brad, but he couldn't find the words. He sat down in a chair and also looked out the window. They stayed this way for what seemed a long time.

"I'm leaving soon."

The suddenness of the unexpected words startled Toby, who had been in deep thought about their problems. "What do you mean, you're leaving?"

"Just what I said. I gave my parents the bad news…it was cancer, by the way…and they want me to come home for treatment. I'm dropping out of school for a while. Hopefully, I'll be able to return after treatment."

Toby was filled with sorrow, anger, and fear, all at the same time. But he had to know what Brad was facing. "What did Dr. Cruz tell you?"

"That I have oral cancer. In fact, I have an aggressive strain of it. As a result, I will require extensive surgery, if I'm going to beat this thing. Before that, I need a thorough work-up that includes all sorts of tests to determine the extent of my problem." Brad suddenly became emotional. With tears in his eyes, he said, "I'm going to win, Toby. I'm going to beat the Big C."

Toby's eyes also filled with tears as he said, "You will. You will." He hurried to Brad, who had gotten up from his chair, and hugged him. With tears flowing freely, he repeated, "You will!"

VIII

A week later, Toby was admitted to the university hospital where, under general anesthesia, he had a complete excision of his mouth lesion. A skin graft was taken from the inner aspect of his left upper arm. A plastic mold was attached to the inside of his mouth, to hold the graft in place. He was discharged the same day.

With his parents at his side, Toby was pushed in a wheelchair to their van. He appreciated that throughout his ordeal, both his mom and dad had been understanding, and not judgmental about his problem with smokeless tobacco and its consequences. They only wanted their son to be well again. They helped in any way they were able. Even though the temperature was not much above freezing, the sun was warm on Toby's body and it felt good to be alive. Toby had a pressure wrapping over the skin graft donor site, and his mouth was plumped out from the plastic mold, but he felt surprisingly little pain. Above all, he had a feeling of relief, since he had—most likely—been cured of his disease. Once in the van and on the way home, his mom, who was driving, began to softly cry, quickly running through tissues.

"Don't cry, Mom. I'm fine. I'll have to be checked frequently for quite a while, but we got it in time."

"I know, Toby, I guess I just feel like crying. I'm grateful that you are going to be well again. I'm also thinking of Brad. Knowing that he is going to lose part of his lower jaw and needs major surgery in his neck, breaks my heart."

"Mine too, Mom. He may also need radiation therapy or chemotherapy, depending on how his cancer responds or spreads. Even under the most ideal circumstances, Brad will probably miss a whole year of school. And as far as bodybuilding is concerned, who knows if and when he can get back into it. I'm praying that everything turns out okay for him. I'm going to visit him on my next vacation."

Mr. Stevenson, sitting in the back seat of the van, had been uncharacteristically quiet that morning. Staring out the side window as the blurred scenery passed by, he scarcely more than muttered, "When will people learn that any form of tobacco can kill?"

Definitions: Toby

Smokeless Tobacco: Tobacco (snuff, chew) that one places in the mouth, usually between the lip and gum. Effects of the tobacco are the result of absorption via the mucosa of the mouth. Often called snuff dipping.

Mucosa: The lining of the mouth that is the equivalent of skin. Mucosa lines other surfaces of the body, such as the gastrointestinal tract.

Ulcer: The loss of the outer (superficial) surface of the skin or mucosa. Commonly called a sore or cut.

Inflammatory Reaction: The body's reaction to irritations such as the products of tobacco. Blood flows into the area bringing the products of healing. Heat, redness, and swelling are part of the inflammatory reaction.

Erythema: Redness associated with the inflammatory reaction. Often, it is a sign of chronic (long-term and/or repeated) irritation.

Leukoplakia (white patch): The buildup of surface cells of the mucosa of the mouth in response to chronic irritation. It may transform into cancer if the irritation is not removed. On the skin such a buildup of cells is often called a callus.

Dysplasia: The change from normal to abnormal cells of the mucosa. Dysplasia can be mild to severe. It may be considered the first stage of cancer.

Cancer (malignant): A general term referring to various types of malignant tumors. In Brad's case cancer occurred when his irritation progressed from the superficial tissues into the deeper tissues of his mouth. Once the deeper tissues are involved, the cancer may spread (metastasize) throughout the body.

Biopsy: The removal of part (incisional biopsy) or all (excisional biopsy) of a lesion (disease process) for evaluation under a microscope in order to obtain a diagnosis.

Graft: The replacement of lost tissue (receptor site) with tissue taken from another part of the body (donor site). It may be skin, mucosa, bone, etc.

Epilogue

By reading this book, you have gained a great deal of knowledge about the profession of oral and maxillofacial surgery. The chapters are not meant to be scientific treatises for those in the profession; rather, they are, hopefully, interesting and informative stories designed to be easily read by most people, especially the younger reader. While the characters and events in the chapters are fictional, their problems are handled as they might be by any modern oral and maxillofacial surgeon. A great deal of medical and surgical information is intertwined with information about such subjects as Meharry Dental School, autism, mountain biking, karate principles, women's softball, and more. The definitions at the end of each chapter are for those who crave more detailed scientific information. I hope it has been an enjoyable learning experience.

Yet, some say, "You have to be sadistic to choose your field of endeavor," or "You must like to hurt people." Those statements are far from the truth. We want to help. We strive to make you feel better. Just ask the woman who has been up all night with a toothache. Ask the diabetic man with swelling caused by an infected tooth that could endanger his life. Ask the person who is so ashamed of her teeth that she places her hand over her mouth when she speaks. Ask the young man who, when questioned why he wanted corrective jaw surgery, declared that he hated to look in a mirror. You get the picture.

One final bit of knowledge about our group which you should be made aware of is that we have a great weakness. We are suckers for thank-you cards, pictures drawn by children about their visit to the dentist, and plates or boxes full of cookies. Give these items to us, and you give us a needed smile in what is usually a very busy day. Just like you, we love to be appreciated.

DRG